KAY'S INCREDIBLE INVENTIONS

A FASCINATING AND FANTASTICALLY FUNNY GUIDE TO THE INVENTIONS THAT CHANGED THE WORLD*

*AND SOME THAT DEFINITELY DIDN'T

ADAM KAY

ILLUSTRATED BY HENRY PAKER

PUFFIN

PUFFIN BOOKS

UK | USA | Canada | Ireland | Australia
India | New Zealand | South Africa

Puffin Books is part of the Penguin Random House group of companies
whose addresses can be found at global.penguinrandomhouse.com

www.penguin.co.uk   www.puffin.co.uk   www.ladybird.co.uk

Penguin
Random House
UK

First published 2023

001

Text copyright © Adam Kay, 2023
Illustrations copyright © Henry Paker, 2023

The brands referred to in this book are trademarks belonging to third parties

The moral right of the author and illustrator has been asserted

Text design by Sophie Stericker
Printed in Great Britain by Clays Ltd, Elcograf S.p.A.

The authorized representative in the EEA is Penguin Random House Ireland,
Morrison Chambers, 32 Nassau Street, Dublin D02 YH68

A CIP catalogue record for this book is available from the British Library

HARDBACK
ISBN: 978–0–241–54078–7

INTERNATIONAL PAPERBACK
ISBN: 978–0–241–54079–4

All correspondence to:
Puffin Books
Penguin Random House Children's
One Embassy Gardens, 8 Viaduct Gardens, London SW11 7BW

MIX
Paper | Supporting
responsible forestry
FSC
www.fsc.org   FSC® C018179

Penguin Random House is committed to a
sustainable future for our business, our readers
and our planet. This book is made from Forest
Stewardship Council® certified paper.

KAY'S
INCREDIBLE
INVENTIONS

This book is dedicated to BUTLERTRON-6000,

my robot butler, for providing a fact check throughout.

It is also dedicated to me, Adam Kay,

the greatest inventor of all time.

⚡ **Fact check - this is completely inaccurate.** ⚡

CONTENTS

INTRODUCTION

Take a look around you. What can you see? Maybe you're in the bedroom – there's a light above you, an alarm clock next to you, a duvet on top of you and a bed underneath you. **⚡ Fact check – readers are statistically more likely to read this book on the toilet. ⚡** Perhaps, like me, you're in the living room – there's a TV, a sofa and a dog being sick into the fireplace. Pippin, stop that!

Did you ever wonder where this stuff around you all came from? No, not from Amazon. I mean, who was the first person to have the amazing idea of making a mobile phone or the annoying idea of building a school? Everything around you that doesn't twinkle in the sky or grow or woof or fart was invented by someone. Like the scientist who invented the microwave by accident, the teenager who came up with the trampoline and the genius behind Smell-O-Vision.

We'll meet the queen who did the first poo on a toilet, hear about the light bulb made out of beard hair and learn why margarine used to be full of maggots. You'll find out why Ancient Greeks wiped their bums on dinner plates, why the Hoover was almost called the Spangle, why the first submarine was built out of leather and grease, how Velcro was invented by a dog, why a whole country was banned from eating sausages so they could make planes and how a movie star invented Wi-Fi (no, it wasn't Zendaya).

I'll also tell you about the most important inventor of all time . . . me. As well as inventing the world's first robot butler, the BUTLERTRON-6000, I am the mega-brain behind Coriander Custard and the Ultimate Underwater Piano, plus over eight thousand other equally brilliant inventions, which you can order directly from my company Adam Kay Genuis Enterprises Limited.

➤ **Fact check – you spelled 'genius' incorrectly.** ➤

If any of that sounds interesting, read on! And if it doesn't, you should put this book down and read some old rubbish instead, like *My Favourite Types of White Paint* by Gerald Humbum. ➤ **Fact check – *My Favourite Types of White Paint* is a superior and much more popular book.** ⚡

PART ONE

AT HOME

LIGHT BULBS, LASAGNE AND LOO ROLL

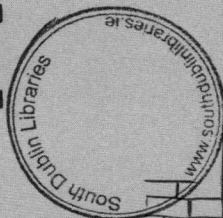

THE
BATHROOM

Let's start with the only room in the house where it's totally fine to sit down and poo – the living room. ⚡ Fact check – the bathroom. ⚡

Attention, any time-travellers reading this book from one hundred years ago: the bathroom is a room in houses these days where people can poo and brush their teeth and have a bath.

Attention, any non-time-travellers reading this book: I had to explain that because the bathroom is a pretty recent invention. In this chapter I'm going to tell you who we have to thank for the fact we don't have to wee in the garden any more. Or who to blame, if you like the idea of weeing in the garden.

Also attention, any time-travellers reading this book from the future. Please can you tell me if my dog, Pippin, will ever stop pooing in the dishwasher? Will the Octopus People of Zaarg be kind to humans when they take over Earth in the year 2185? Oh, and what are next week's lottery numbers, please?

THE NUMBERS ARE 2, 15 . . . UMM . . . I REALLY SHOULD HAVE WRITTEN THEM DOWN. SORRY.

STRANGE BUT POO

Toilets have been around for as long as bums have, which is ages. Poos and wees don't smell great – unless you're Pippin, in which case they're a delicious snack. And that's why people have always been keen to get them away from their houses. In Ancient Rome, two thousand years ago, people would visit public toilets. So far so good. About twenty people would sit on a long bench with holes cut into it and poo into a pit underneath, while they chatted to each other about where they were going on their holidays and where they bought their togas. So far so terrible.

Until about six hundred years ago, people pooed out of the window quite a lot – either straight out, or into a pooey pot first. If you lived in a castle, then it would splat into the moat, which the fishes can't have liked very much. But if you lived in a tall building, it would just land on the street. Luckily for the people walking underneath, umbrellas had already been invented. (My lawyer, Nigel, has asked me to mention that it is both illegal and disgusting to poo out of your window.)

A ROYAL FLUSH

It's difficult to know what present to buy for a king or a queen. A diamond necklace for their corgi, maybe? A massive rubber duck to put in their moat? Four hundred years ago, John Harington had this exact problem – it was made even worse by the fact that Elizabeth I was his godmother, so he really didn't want to mess it up. Apart from anything else, she was a big fan of chopping people's heads off.

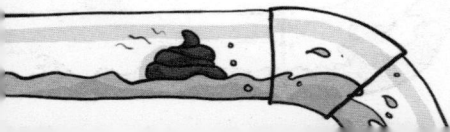

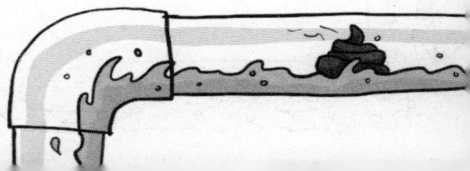

But John had a great idea. He'd just invented the first-ever toilet you could flush and named it the Ajax. He thought old Queenie would really like a massive toilet as her present, so he had one built in Richmond Palace. Elizabeth loved her amazing new toilet, and she did some of her favourite poos in it. ⚡ **Fact check – there is no evidence that Queen Elizabeth I had any favourite poos.** ⚡

UNIVERSAL RATING FOR INVENTION NAME (URIN)

8/10

VERY FUTURISTIC

I WONDER WHY HER DRESS IS SO BIG?

ROOM FOR IMPOOVEMENT

The palace privy didn't really become popular (except with Queenie) because people didn't have water pipes going into their houses back then, so after every royal wee a bunch of servants would have to go to a well, fill up three big buckets, then carry them upstairs and splash them into the toilet. I don't know about you, but I don't have three servants – I've only got a robot butler and he doesn't obey any of my instructions. **Fact check – I have actually obeyed almost four of your instructions in the last ten years.**

Before long, the toilet situation became quite urgent, because people were getting ill from all the brown stuff that was plopping straight into the rivers and then their drinking water. First on the to-do list was getting some sewers – great big underground pipes to safely whizz away everything that came out of people's toilets. And then, gradually, lots of improvements to toilet technology started dribbling through like drops of diarrhoea. One of the biggest improvements was thanks to a man who had absolutely nothing remotely funny about his name, Thomas Crapper.

In 1880, the toilets in people's houses absolutely stank. I know, I know – toilets these days don't exactly smell like freshly baked bread . . . but this was a thousand times worse. The stink fumes from the sewer would rise straight up through people's toilets and make the bathroom smell like Godzilla had been in there after eating a cauldron of chilli. The solution came from Thomas Crapper, who absolutely didn't have anything named after him – I don't know what on earth you're thinking. His invention was called a U-bend, and if you look behind your toilet you'll see that we still use them today. If the pipe that goes out of the toilet is shaped like the letter U, the smelly fart-gases can't escape from the sewers and knock people unconscious. If you look inside your toilet – no, not in the toilet bowl; I mean inside the bit on top where you flush it – you'll see a floating ball on a stick. This is called a ballcock and is another one of Tommo's inventions, which stops too much water flowing in and causing a flood. (My lawyer, Nigel, has advised me to mention that you should get an adult to help if you want to look inside your toilet. He's far too busy to be sued because people have broken thousands of toilets after reading this book.)

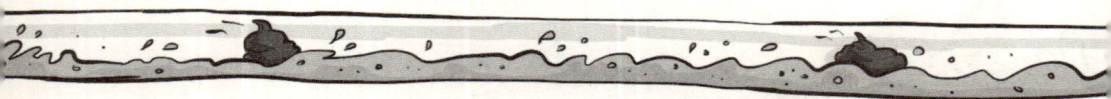

Now, let's ask my robot butler to activate his lie detector to see if you can guess which Thomas Crapper fact is a total pork pie. (That's rhyming slang for 'lie'.)

ROBOT BUTLER'S

LIE DETECTOR

1. Thomas Crapper opened the world's first shop where you could buy a toilet.

2. He sadly died by drowning in a toilet.

3. Hundreds of members of the royal family have put their bottoms on toilets he designed.

4. If you visit Westminster Abbey in London and look down, you can see lots of manhole covers with Thomas's name on from when he fixed the drains there.

5. Thomas Crapper toilets are still being made today.

2. Mr Crapper's death did not occur in a bathroom.

BACK TO THE POOTURE

If you're happy to spend more money on a toilet than I spent on my car, then these days you can get one with the following features:

- Adjustable height

- Bluetooth speakers (I'd have called them Poo-tooth speakers, personally)

- A heated seat

- A jet of water to wash your bum

- Different coloured lights

- A bum dryer

- Time travel

Fact check – no toilet is currently capable of time travel.

TINA AND HER TERRIFIC TIME-TRAVELLING TOILET!

STOP! YOU'RE ABOUT TO GET POO ON YOUR HANDS!

Maybe it's just me, but if I had spent ten thousand pounds on something, then I probably wouldn't want to poo on it.

LOO GOTTA ROLL WITH IT

Loo roll was invented by a woman called Lou Roll, who – ⚡ **Fact check – my quality-control module suggests you should start this section again.** ⚡

OK. Loo roll has only been used in the Western world for about one hundred and thirty years, and I'm fairly sure you'll agree that people have been pooing for longer than that. So what was everyone using until then? Well, after the Ancient Romans had finished sitting on their big long toilet benches chatting to each other, they would grab a sponge on a stick and clean themselves up. And then the next person would take the same sponge on a stick and use it. And so would the next person . . . and the next person . . . Now is probably a good time to let out an enormous sigh of relief that you don't live in Ancient Rome.

NO! NOT THE STINKY STICK!

And in Ancient Greece, if there was someone you really didn't like, you could get their name written on a dinner plate and then use that every time you went to the bathroom. When *pootery* went out of fashion, people would use pretty much anything they could get their hands on – leaves, grass, animal skins and even corn on the cob, which doesn't sound like it would be particularly comfortable.

I THINK YOU'RE SUPPOSED TO WAIT TILL I'M DEAD.

In China, they sussed out the whole toilet paper idea about seven hundred years ago, but it took a lot longer for it to reach Europe or America. Back then, the Chinese would use sheets of paper in a box, a bit like printer paper. I just hope nobody got a paper cut . . . The first actual roll arrived in 1857 in America, and I'm pretty sure people started arguing about whether you should fold it or scrunch it up to use it about two days later. I'm #TeamScrunch. In 1952, companies started making coloured toilet paper, so it could match your bathroom or your socks or whatever. The brown colour didn't sell very well, for some reason.

YOU DO THE BATH

Until one hundred and fifty years ago, houses didn't have bathrooms. You basically had three options. Option one: go to a public bathhouse, which was like a big naked swimming pool with soap and sponges. Option two: keep a metal bath in your garden, and once a week carry it inside, put it next to the fire and fill it up with loads of buckets of water that you've heated up. Because this was a massive faff, everyone would take turns using the bath one after the other, from the oldest member of the family to the youngest. So you'd better hope that Grandpa doesn't like weeing in the bath. By the time the youngest person got into their disgusting weekly bath, the water would be so brown and murky that they couldn't even see their own toes. Option number three was to just stay stinky. This is Pippin's preferred method. ➤ **Fact check – and yours.** ➤

When people started getting water delivered directly to their houses (by pipes, I mean, not the postman), they finally bought baths with taps. These were often clawfoot baths with those funny little feet that make it look like a lion got cursed by a witch and turned into a bath. We used to have a bath like that, but Pippin thought it was an actual animal and would woof at it so much that we eventually replaced it with a normal tub. I don't think she's the cleverest dog in the world, to be honest.

IT'S BARF TIME!

19

THE SHOWER AND THE GLORY

William Feetham was famous for having feet made out of ham. ➤ **Fact check - William Feetham was famous for inventing the first-ever shower.** ➤ The first-ever shower looked more like a big wooden wardrobe that you stood inside. Instead of fiddling with dials and knobs, you would just pull a big chain and a trap door would open above you and a huge bucket of water would fall straight onto your head in one go. Here's a diagram of how it worked, plus a picture of a Santa drinking some Fanta if you'd prefer to look at that instead.

Because one splosh isn't enough for a shower, you'd then turn a big handle, which would pump the water back above your head. And then repeat. This did mean that every time you pulled the chain, the water was colder and dirtier than last time, which wasn't ideal. Also, it meant you couldn't wee in the shower, because soon afterwards it would be landing on your head. I should point out that I never wee in the shower.

➤ **Fact check – you actually always –** ➤ Oh dear, we've run out of space.

MEMBERS OF THE CABINET

Let's have a quick look at the various lotions and potions and gadgets and gizmos inside your bathroom cabinet.

TOOTHBRUSH

Humans have known for thousands of years that if they didn't clean their teeth, then they'd crumble like chalk. Their teeth would crumble, I mean, not their whole bodies. Even inside the pyramids, archaeologists found twigs that pharaohs used to brush their teeth. (Although

ELECTRIC NOSE-PICKER

WORST CHEESE EVER!

SPONGE

IMFLOSSIBLE NEVER-ENDING FLOSS

COTTON BUDDIES

I guess they might have just been twigs that fell in there from a tree.) The first toothbrush that you'd recognize today was made by an English bloke called William Addis. He had been sent to prison in 1770 for starting a riot, and one night he sneaked a bone from his dinner into his cell, then stuck a load of bristles from a pig into

one end of it, and used it to clean his teeth. Nope, I've got no idea why there was a pig hanging around in prison either.

When Wills got out, he opened a toothbrush factory – and that company, Wisdom, is still producing millions of toothbrushes over two hundred years later. I'm pretty sure they don't make them out of bones and pig-bum-bristles any more. ➤ **Fact check – you are correct, for the first time in this book.** ✚

The first electric toothbrush was invented around seventy years ago, in 1954, by an excellently named scientist: Dr Philippe Guy Woog. He called his invention the Broxodent, which was basically French for BrushyTeeth. I might start calling toothbrushes that now, actually.

UNIVERSAL RATING FOR INVENTION NAME (URIN)
9/10
GREAT IMPROVEMENT.

TOOTHPASTE

If you've ever moaned about using a BrushyTeeth, think how much moanier you'd be if you'd lived in Ancient Egypt. You would be cleaning your teeth with a mixture of crushed-up bones, old eggshells, wee, hooves and spices. Even two hundred years ago, things weren't much better – people would use a mixture of soap and crushed-up toast on their BrushyTeeth. The first toothpaste that looks like the stuff we use today came out in the 1870s, thanks to a dentist who was named after two cities, Washington Sheffield. His toothpaste was white and minty, came in a tube and worked a lot better – everyone was much happier. Except perhaps the tooth fairy.

DEODORANT

It's time to stink about, sorry, *think* about body odour. Back when people lived in caves, nobody really cared about a bit of armpit stench, because they mostly concentrated on not getting eaten by some horrible creature with loads of claws and fangs. In fact, a bit of a pong might have helped them – what animal wants to eat someone who smells like a glass of milk that's been left in a hot car for ten years? The Ancient Egyptians were the first people to come up with a kind of deodorant: they smeared porridge on their armpits. Maybe they thought it was better to pong of porridge than stink of sweat?

I MIGHT SKIP PORRIDGE TODAY . . .

Then about one hundred years ago a woman called Edna Murphey invented something that actually worked! Her dad was a doctor who wanted to stop surgeons getting sweaty hands and accidentally dropping their scalpels into people's hearts. I would like to reassure any former patients of mine reading this that I never accidentally dropped a scalpel into their heart.

Together, Edna and her dad developed a liquid called Odorono, which worked, and she started selling it to anyone who had an armpit. It wasn't totally perfect. It was so acidic that it burned through people's clothes (including one woman's wedding dress – oops). But it was the start of an industry that's now worth more than ten billion pounds a year, and means that people don't faint from the stink every time they get on a busy bus.

Do you want to learn about soap? Or would you lather not? ➤ **Fact check – my joke-assessment module informs me that this has a 3% humour level.** ➤

People have been making soap for about five thousand years. I think my Great Aunt Prunella still has some of that original stuff in her bathroom. Early soaps were made out of a mixture of animal fat and ash, which sounds like it would get you even dirtier. Early *soups* were made out of leek and potato. If you took a microscope and peered at a molecule of soap, you'd see that it looks a bit like a tadpole – one end of it loves water and the other end of it loves oil. Germs and dirt get mixed in with the natural oils on your skin, and when you put soap on, the oil-loving bit joins to it. And then, when you wash your hands, the water-loving bit of soap jumps over to the water, taking the germs and grease and gunk with it. Here's another diagram, and an alternative picture of a yeti eating spaghetti.

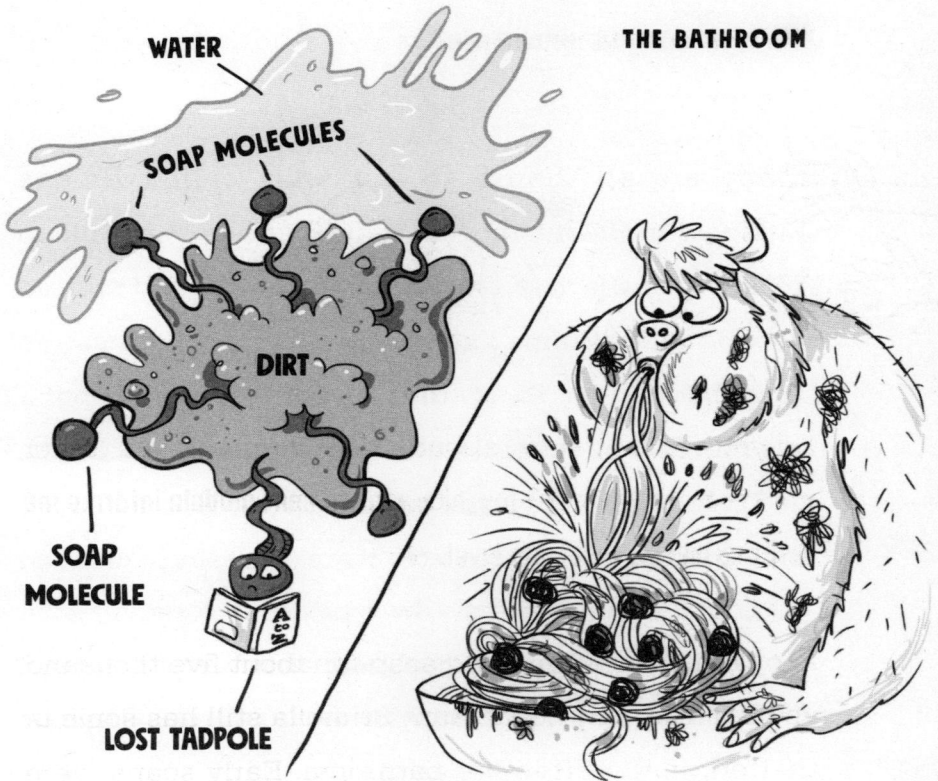

WATER

SOAP MOLECULES

DIRT

SOAP MOLECULE

LOST TADPOLE

FACE CREAM

If you wanted to take care of your skin in Ancient Rome, there were various options available, such as smearing your face with goose fat, crocodile poo, gladiator sweat, urine or poisonous lead. I think I'd have just splashed my face with water, to be honest. Things got slightly better four hundred years ago, when face creams were made of things like lemon juice, egg whites and rhubarb – although that does sound more like a recipe for a meringue to me. But face creams as we know them today are all thanks to a woman called Helena Rubinstein. Helena moved from Poland to Australia in 1896, when she was twenty-six years old, and started making face creams like the kind her mother used to make back home. To say these creams became popular is a bit of an understatement – not long afterwards she had sold so many that she became the richest woman in the entire world! I put on a face cream every single morning without fail, and that's why I look like a movie star.

⚡ Fact check – my image-assessment module informs me that this is correct. You have a strong visual similarity to King Kong. ⚡

MAKE-UP

All I can say is be very, very grateful that we live when we do. It means you avoided Ancient Egyptian lipstick made from squished-up insects, fourteenth-century Italian eye drops that would make you vomit, and eighteenth-century lead foundation that would cause your eyes to turn red and your teeth to fall out. Although I should warn you that some perfumes today do still use whale

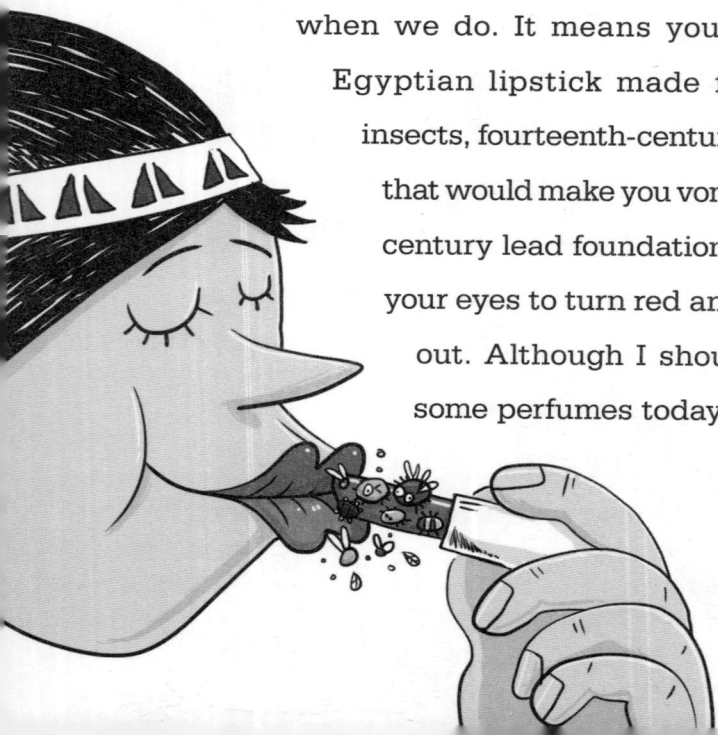

vomit. ⚡ **Fact check - this is correct.** ⚡ That's why it's always best to read the label . . .

VASELINE

I don't know if you've ever used Vaseline for dry skin or scrapes and scratches. It was invented in 1872 by Robert Chesebrough, who made it out of petrol. RobCheese thought it was the cure for everything and when he got a chest infection he would cover himself in Vaseline from head to toe. He also ate a spoonful of it every day because he thought it was good for you. Yuck. Although he did live until he was ninety-six, so maybe it worked after all. (My lawyer, Nigel, has asked me to point out that you should never eat Vaseline because it tastes absolutely disgusting and it definitely isn't good for you.)

ACHOOOOOOOOOD

VASELINE

TRUE OR POO?

BATHROOM MIRRORS USED TO COME FROM THE INSIDE OF A VOLCANO.

TRUE These days mirrors are made of glass with some shiny stuff sprayed onto the back, but what did you do thousands of years ago if you wanted to check out your cave quiff? Well, unless you had a pond you could spy your reflection in, you would polish up a bit of rock and look in that. The shiniest rock was called obsidian – and the only problem was that you had to get it from a volcano. I hope somebody had invented oven gloves.

AAAGH, THIS HURTS SO MUCH! BUT MY TEETH LOOK AMAZING.

A BATH USES MORE WATER THAN A SHOWER.

TRUE It's important to think about how much water we're using – for the environment, and also to save money. Unless you sing the complete works of Adele every time you take a shower or get in a bath the size of a bucket, you'll use less water by having a shower. I save water by using the bath straight after Pippin has been in. ⚡ Fact check - this explains your unusual odour. ⚡

THERE IS A TOILET-BASED THEME PARK IN SOUTH LONDON.

POO Sorry if you were planning a day out. But if you're prepared to travel a bit further, you can visit the world's only toilet-based theme park in South Korea. It was founded by a man called Sim Jae-duck, who was born in a toilet, and devoted his entire life to toilets. He was even known as 'Mr Toilet' (although I think 'Toilet Duck' would have been a better nickname). He lived in a house shaped like a giant toilet, and in 2012 he turned it into a theme park full of hundreds of toilets. You could say . . . he was absolutely potty about them. That's an excellent joke, by the way. ⚡ Fact check - my joke-assessment module informs me that this has a 1% humour level. ⚡

ADAM'S ANSWERS

HOW MANY PHONES GET DROPPED INTO TOILETS EACH YEAR?

In the UK about two million people accidentally drop their phones into the toilet every year. If you do this yourself, then dry it out immediately by putting it in a bowl full of rice. And if you accidentally drop a load of rice into the toilet, then dry it out by putting it in a bowl of iPhones. (My lawyer, Nigel, has asked me to point out that you should do neither of these things.) If you know someone who takes their phone into the bathroom, you should tell them that it becomes absolutely covered in all sorts of horrible germs . . . and then they hold it up to their mouth. They might as well smear a poo on their face.

WHY IS THE CITY OF BATH CALLED BATH?

If you go to the very centre of Bath, you'll see there's a giant plughole called the Mighty Drain, where all the water flows away after it's been raining. And on the side of the town hall there's a great big tap that all the residents use for drinking water. There's even an ancient yellow sculpture on top of nearby Solsbury Hill called Ye Olde Rubber Duckie and – **Fact check - every single word of this is incorrect.** I'll try again. Bath is called Bath because there are natural jets of hot water that come from really deep inside the earth, so Romans would go there to swim because it was like a lovely hot . . . bath! Sadly there isn't a city called Shower where there's always loads of nice warm rain.

HOW MANY OF US WASH OUR HANDS AFTER GOING TO THE TOILET?

I'm pleased to say that three out of four people wash their hands afterwards. But that means I'm also totally disgusted to say that one in four people don't. I'm just thinking of all the people I've ever shaken hands with in my life – a quarter of their hands had wee on them. I'm suddenly not feeling particularly hungry.

DEATHVENTIONS

It's a dangerous business being an inventor, especially if you're a totally useless one. Here are some inventors who ended up splatted, exploded or otherwise deaded by their own discoveries. In case it isn't clear, the moral is . . . DON'T TRY THIS AT HOME! (My lawyer, Nigel, would like me to emphasize this very strongly.)

BIRD SUIT

In around 1010 (that's the year 1010, not ten past ten in the morning), a man called Ismail ibn Hammad al-Jawhari decided that flying didn't look particularly difficult. After all, birds and wasps could do it, and they're idiots. So he climbed onto the top of a tall building, strapped some wings made of wood and feathers onto his arms, jumped off the roof and started flapping. What do you think happened next? Well, I can tell you what didn't happen – he didn't fly.

NOT AS EASY AS IT LOOKS, IS IT?

BED PULLEY

Thomas Midgley was an inventor all his life, coming up with loads of useful things, from a new kind of petrol to colder fridges and even squirty cream. When he developed an illness that meant he lost strength in his legs, he invented a clever system of ropes and pulleys so he could get out of bed easily every morning. Well, it wasn't such a clever system – one day it strangled him to death. Oops.

NEWSPAPER PRESS

About two hundred years ago, a man called William Bullock invented the fastest-ever way to print newspapers. His machine took huge rolls of paper, printed on both sides of it, folded it up into newspapers and cut them down to the right size. Unfortunately, it also did one other thing – it sucked William inside and squished him as thin as a newspaper.

ANYTHING IN THE POST TODAY?

BAD NEWS, I'M AFRAID – WILLIAM'S DEAD.

STEAM-POWERED BICYCLE

In 1896, Sylvester Roper invented the steam-powered bicycle. If you're wondering why you don't see any steam-powered bicycles on the road today, it's because when he was trying it out one day, things got a bit wobbly and he fell off. RIP.

PARACHUTE SUIT

Franz Reichelt was a tailor who thought he could improve on those great big parachutes in backpacks that pilots wore. He came up with a magical jacket that could transform into a parachute if your plane decided to fall out of the sky unexpectedly. And to prove it worked, in 1912 he put his suit on, invited all the world's press to watch him and jumped off the top of the Eiffel Tower in Paris, which was the tallest building in the world back then. Unfortunately for Franz, he's in the section about people who died because of their inventions, so you can probably guess what happened :-(

ON THE PLUS SIDE, YOU'LL BE IN A BOOK ONE DAY.

NEW FROM

ADAM KAY GENUIS ENTERPRISES LIMITED

ADAM'S HANDY HAIRCUT HAT

Fed up with wasting time going to the hairdresser's when you could be at home reading books by Adam Kay? Pop this hat on and, five minutes later, your hair will be cut to the smart and sophisticated style of your choosing.*

Only £5,244.99 (excludes batteries and blades)
*Please note that the only style currently possible is 'looks like you've been attacked by a tiger'.

THE
BEDROOM

Ahh, the bedroom! A place to have a lovely sleep, play video games and read some books by record-breaking author Adam Kay. **⚡ Fact check – your world record was for most spelling mistakes in one book. ⚡** I'll have you know that my spelling is exalent.

The average person spends thirty-three years of their life in their bedroom, so it's time to learn about all the people who've made this possible – from Captain Willy Wardrobe, inventor of the wardrobe, to Penelope and Patricia Pillow, the twins who discovered the pillow. **⚡ Fact check – overload. Too many incorrect facts. ⚡**

THE LIE-IN KING

Beds have been around for as long as humans have – it didn't take long for people to notice that it's not very comfortable to lie down on a big load of rocks. The oldest mattresses that have been found are two hundred thousand years old (my Great Aunt Prunella still uses those) and were big enough for entire families to sleep on. Bad news if you've got a dad who smells of turnips and the world's fartiest niece. **→ Fact check – I calculate your smell to be twenty-three times worse than any other member of your family.** ✦

In Ancient Egypt, the pharaohs would sleep on solid gold beds, which don't sound great for your back, if you ask me. And three hundred years ago, the French King Louis XIV (that's Louis the Fourteenth, his name wasn't pronounced Louis Sieve) loved beds so much he had 413 different ones, and he would sleep in a new one every night. He would even carry out important meetings from his bed, and if he fell asleep during one of those meetings, it was considered a great honour. I might start doing that in boring meetings with my publishers. I hope they're not reading this.

In 1968, Charles Hall was given a university project to design some really comfortable furniture. His idea was simple – an armchair full of jelly. It was a disaster. The chair was so heavy it took six people to lift it. Also, it wasn't particularly comfortable, and eventually the jelly went mouldy and it smelled worse than a skunk's armpit. (Almost as bad as your bedroom.) So he had another go – this time he made a mattress that was filled with water. The waterbed was a huge success, and soon one in five homes in America had one. (My lawyer, Nigel, has asked me to point out that it's important not to buy a waterbed if you have a pet hedgehog.)

The zzz-based invention that I've always wanted is the Murphy Bed. It was invented over one hundred years ago by a bloke called William Murphy who lived in an absolutely tiny single room in New York where there was barely space for a bed. So he designed one that folded up and disappeared into the wall during the day! A brilliant idea, and only a few people have been splatted to death by them. Hmm, maybe I don't want one so much now.

CHIME TO WAKE UP

Everyone hates alarm clocks, so let's find out who's to blame for inventing them, then we can turn up at their houses at 2 a.m. and honk a foghorn in their faces as revenge. In Ancient Greece, two thousand years ago, there was a famous philosopher called Plato. Being a philosopher means you spend your day thinking about things, which seems like quite an easy job to me. He invented an alarm clock that dripped water from one cup to another cup. In the morning, when the water had reached a certain level, it made a whistle blow to get him up so he could start doing all his philosophizing. A few years later, Plato went on to invent plates. ➤ **Fact check - Plato did not invent plates, but he was one of the most famous writers and philosophers in the history of mankind.** ➤

Alarm clocks as we know them today weren't invented until 1787, by an American man called Levi Hutchins, who worked as a clock repairer and had ten children. To be honest, I'm surprised one of those ten children didn't wake him up every morning. There were two slight problems with his invention. Firstly, it didn't fit on your bedside table, because it was the size of a microwave.

Well, it would have been, if microwaves had been invented. Secondly, it could only wake you up at 4 a.m., which sounds like a total nightmare.

Over the years, alarm clocks became adjustable, but they were still far too expensive for most people to afford. And so, even until about fifty years ago, there was a job called a knocker-upper. You would pay a knocker-upper to turn up outside your house every day to bang on your bedroom window with a long stick or throw stones at your house, to wake you up in time for work.

Pippin wakes me up every morning by licking my face – I hope she doesn't expect to get paid for that . . .

CRUMBS!

There's not much that's better than breakfast in bed –
though not when your robot butler accidentally mixes
your orange juice with your coffee and leaves loads of
shell in your scrambled egg. **Fact check - that is never
an accident.**

But I wasn't the first person to come up with the idea of
breakfast being delivered straight to your pillow. That
particular genius was called Sarah Guppy, and she lived
in England about two hundred years ago. She was an

extremely successful inventor, who had made a fortune by working out how to remove barnacles from boats and build brilliant bridges. She clearly specialized in things that began with the letter B. And next on her list was . . . breakfast. Her amazing steam-powered gadget would make you a cup of tea, cook an egg, then heat up some toast and bacon and put it on a nice warm plate for you. It's a disgrace that these aren't in every single bedroom in the country.

Now, let's ask my robot butler to activate his lie detector to see if you can guess which Sarah Guppy fact is a total dragonfly. (That's probably rhyming slang for 'lie'.)
➤ Fact check - it is not. ➤

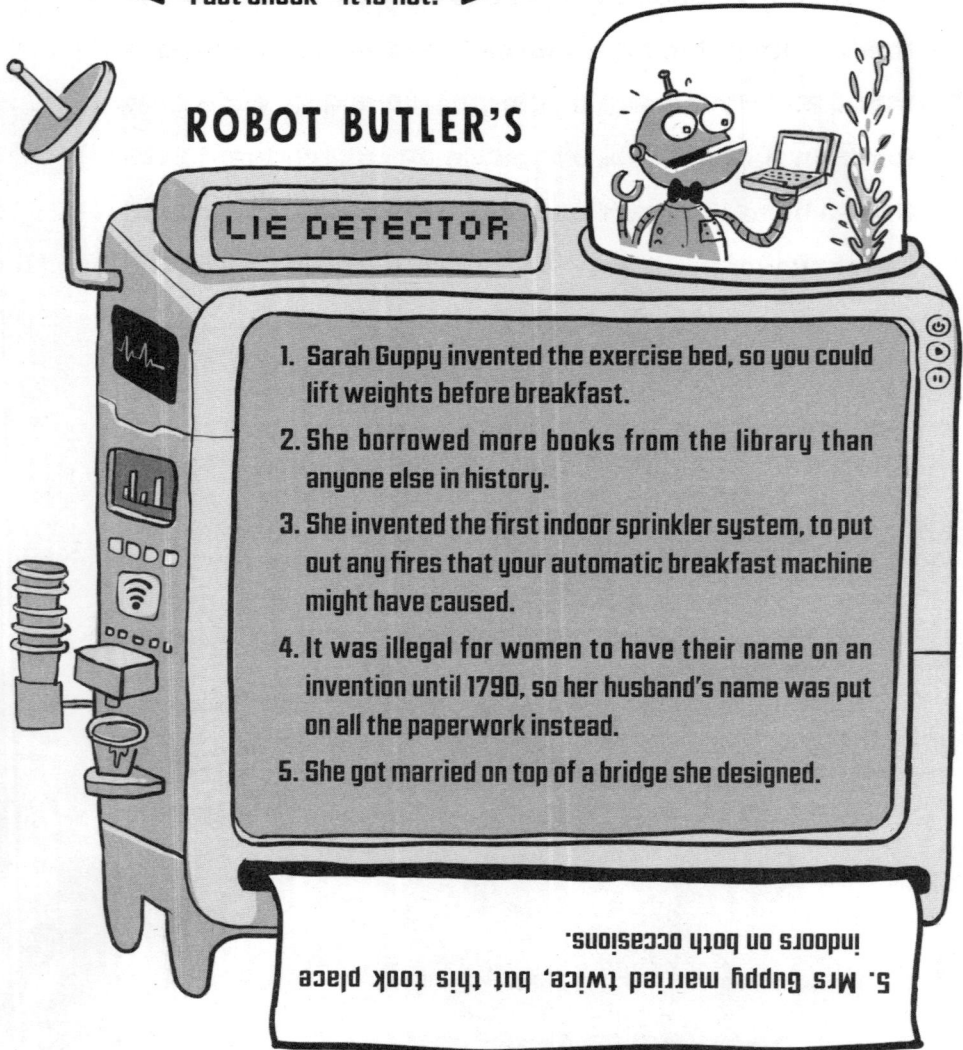

ROBOT BUTLER'S

LIE DETECTOR

1. Sarah Guppy invented the exercise bed, so you could lift weights before breakfast.
2. She borrowed more books from the library than anyone else in history.
3. She invented the first indoor sprinkler system, to put out any fires that your automatic breakfast machine might have caused.
4. It was illegal for women to have their name on an invention until 1790, so her husband's name was put on all the paperwork instead.
5. She got married on top of a bridge she designed.

5. Mrs Guppy married twice, but this took place indoors on both occasions.

HAIR WE GO

Just got out the bath? Want to dry your hair? Accidentally time-travelled to the year 1900? Well, you can't use the hairdryer, I'm afraid. Back then they were massive machines the size of a filing cabinet that were only found in hairdresser's. If you time-travel forward from there a bit to 1920, portable hairdryers had been invented but they were still very heavy, complicated to use and quite often caught fire or electrocuted people. Maybe you should just use a towel instead.

And how about if you want to straighten your hair or give it a bit of a curl? Well, you've got Marjorie Joyner to thank for that – her invention in 1928, which led to the hair straighteners and curlers that we use today, made her the first African American woman to ever receive a patent, the official recognition that an invention is yours. Another Black inventor who made a big difference to haircare was Garrett Morgan, who invented a hair-straightening cream totally by accident in 1905. He was in the middle of working on a type of polish for sewing-machine needles when he stroked a dog, and its hair went from wavy to completely straight. He suddenly realized that he'd invented a whole new type of hair product. I wonder what Pippin would look like with straight hair . . . **Fact check – my image-generation module informs me that she would look even more ridiculous.** Garrett went on to invent loads of other things: a mask for firefighters to wear so they didn't breathe in smoke, much safer traffic lights, and even a cigarette that could put itself out to prevent fires. (My lawyer, Nigel, has asked me to point out that smoking is extremely bad for you, and that he hopes that you know this already.)

PIPPIN'S HAIRSTYLES
OVER THE YEARS

STANDARD PIPPIN

SLICK PIPPIN

PUNK PIPPIN

POODLE PIPPIN

FLOWING PIPPIN

TOPIARY PIPPIN

GET ME A TAILOR, SWIFT!

Where would we be without clothes? Well, we'd probably get arrested every time we went to the supermarket. But who are the brilliant inventors who took us from wearing sabre-toothed-tiger togas to where we are today: all wearing Adam's Rotating Raincoats – only £2,642.99 from Adam Kay's Genuis Enterprises Limited? **Fact check – none of these have been sold yet.**

SYNTHETIC DYE

For thousands of years, the only way for clothes makers to put a bit of colour in your cardigans was by crushing up flowers (awwww!) or by crushing up insects (aaaagh!). This was expensive, made it difficult to keep the colour the same across your outfit, and also meant that it would get a lot fainter every time you washed it. Then, one day in 1856, a university student called William Perkin was doing an experiment to try to make a medicine to treat a disease called malaria. He got a D minus in this experiment, because all he managed to make was a weird black goo. But he noticed that anything this goo touched turned into a lovely purple colour. He even invented a brand-new name for this colour: janglebangle. ➤ **Fact check - he actually named this colour 'mauve'.** ➤ Willyperks set up a factory next to a river, making different colours of dye using extracts of coal. The colour of the river would change all the time, depending on whether he was making green, black, pink or purple dyes that day – I hope that no fish *dyed*.

URIN SCORE
5/10
TOO MANY VOWELS.

I went to the same university as William and, when I finished my degree, the special cloak I wore at the graduation ceremony was janglebangle-coloured, in memory of his brilliant discovery. ➤ **Fact check - this is true; however, you got an extremely low mark in your -** ➤ We'd better look at some other things now.

VELCRO

Pippin needs a bath whenever she comes back from a walk. Any time I'm washing off all the mud and fox poo and bits of plants that she's covered with, I think things like 'I should have got a goldfish instead' and 'Why isn't my robot butler doing this for me?' **→ Fact check – because it's disgusting. ←** In 1941, a man called George de Mestral was pulling a prickly flower head out of the hair of his dog Milka when he thought to himself, *Oh! This could be a new way of doing up clothes!* He realized that if he had one strip of material with loads of tiny hooks on it (like the flower head) and one strip of material with loads of tiny loops on it (like the dog hair), then they would stick together.

He named it Velcro – from the words '*vel*oure' and '*crochet*', meaning velvet hook – and this weird scratchy sticky stuff became popular in clothes everywhere from ski slopes to space! I used to have a lot of Velcro shoes, because I couldn't tie my shoelaces until I was nine. ➤ **Fact check – you were twenty-three.** ⚡

BULLETPROOF VESTS

For thousands of years, soldiers have worn special outfits to keep them safe from enemies with spiky weapons, but the problem with things like chainmail and armour was that they were made of thick, heavy metal, which meant they could move about as fast as a slug with a sore tail. That was until Stephanie Kwolek came along. She was born in Poland, then moved to America in 1946, where she worked for a chemical company called DuPont. She invented a fibre so strong that if you knitted it into a jumper, bullets couldn't go through it! It was called poly-azanediyl-1,4-phenyleneaz-anediylterephthaloyl.

URIN SCORE
2/10
ABSOLUTELY ABYSMAL.

BULLY-PROOF JUMPER

NICE JUMPER!

The company wanted a snappier name for it, so they called it Kevlar after Stephanie's uncle, Kevin Largebum. ➤ **Fact check – it was called Kevlar because the company thought it sounded good.** ➤ As well as being used by soldiers, Kevlar is also used for *(deep breath)* car tyres, protective clothing for motorcyclists, tennis rackets, sails for boats, the tops of drums, mobile phones, violin strings, hockey sticks . . . and about two hundred other things. But Stephanie didn't stop there – she was also involved in the discovery of Nomex, a fireproof material that pilots and firefighters wear, and Lycra, a stretchy material that cyclists and Spider-Man wear.

ZIPS

The first zip went on sale in 1905 and was designed by Whitcomb Judson, although he called it the C-Curity Clasp Locker. These days half of all zips in the world are made by a company called YKK – if you look at a zip, it probably has those initials on it. If you took every zip they make in a year and put them all in a line, it would wrap round the world 150 times. But please don't do that, because then loads of people's trousers would fall down.

URIN SCORE
4/10
NOT VERY CATCHY, BUT BETTER THAN POLY-AZANEDYL-1, 4-PHENYLENE-AZANEDIYL-TEREPHTHALOYL.

BYE!

SEE YOU SOON!

MISSING YOU ALREADY!

CATCH YOU LATER!

TRUE OR POO?

YOUR MATTRESS GETS LIGHTER OVER TIME.

POO   Your mattress actually gets a lot heavier – doubling in weight in ten years. Want to know why? I'll warn you – it's slightly disgusting . . . Well, tough luck, I'm going to tell you anyway. Your mattress fills up with dead skin cells, sweat and up to ten million dust mites – tiny little eight-legged creepy-crawlies that just love eating flakes of human skin. Sleep well!

IT WAS CONSIDERED GOOD LUCK TO HANG A KNIFE OVER A BABY'S COT.

TRUE   I know what you're thinking – this sounds like it could be *extremely* bad luck for the baby, but hundreds of years ago people believed that dangling knives and scissors over babies would scare off evil spirits.

THE RECORD SPEED FOR MAKING A BED IS FORTY-TWO SECONDS.

POO Forty-two long seconds?! Not for expert bedmakers like nurses Sharon Stringer and Michelle Benkel in London in 1993. It only took them FOURTEEN SECONDS to make up a bed with three sheets, two blankets and a pillowcase. So there's no excuse for taking ages the next time you're asked to change the bedsheets.

THAT'S NOTHING. IT ONLY TOOK ME TWELVE SECONDS TO LAY THE TABLE.

ADAM'S ANSWERS

WHY IS DENIM CALLED DENIM?

Denim was first made in a place in France called Nimes. Except the 'i' in Nimes has got a little hat on it. I'm not sure how you do that. Nimes. Nope. Nimes Nimes Nimes. Grr. Nìmes. No, that's not it. Nīmes. Blimey, no, that's wrong. Nïmes. Aargh! Anyway, denim literally means 'from Nimes' – 'de Nimes' in French. ➤ **Fact check - Nîmes.** ⚡

WHAT WERE THE FIRST PILLOWS MADE OUT OF?

You're probably thinking it would be something soft and squidgy, like grass or mammoth poo. In fact, the first pillows were used in Ancient China, and were made out of slightly non-comfy materials such as porcelain or bronze. At least they'd keep you cool at night, I guess.

BETTER FLUFF MY PILLOW.

HOW MIGHT YOUR CLOTHES SAVE YOUR LIFE?

Well, they could stop you getting frostbite if you're up a mountain. And you'd be glad to be wearing trousers if you fell over when you were roller-skating. Oh, and if you count a parachute as clothing, they're pretty handy if you've just jumped out of a plane. ➤ **Fact check – you have forgotten to mention smart textiles.** ➤ I was just about to mention them, actually. ➤ **Fact check – my lie detector informs me you are not telling the truth.** ➤ Smart textiles (see!) aren't jeans that can do your geography homework; they're clothes that can tell if someone has become unwell. For example, a T-shirt that knows if someone's heart has an irregular beat, socks that check whether a baby is breathing properly, and pants that notice if your bum has fallen off. ➤ **Fact check – there's no such thing as pants that –** ➤ Yeah, yeah, yeah. In a few years' time, when Earth has been taken over by the Octopus People of Zaarg, your clothes will even be able to treat medical conditions. Hopefully including injuries from tentacles.

OOPSVENTIONS

Accidents are normally bad things. Like the time my robot butler dropped loads of oil on the kitchen floor, then I slid on it and landed head first in the dishwasher. ➤ **Fact check - that was not an accident.** ➤ But inventing is all about trying new things, and sometimes that means you discover something amazing totally by accident. Like when I misread a recipe and put marshmallows in my soup, and discovered that marshmallow soup is totally delicious.

THE MICROWAVE

In the 1940s, a scientist called Percy Spencer was working on a new type of radar, to spot submarines underwater. All of a sudden he noticed that the chocolate bar in his pocket, which he'd been saving for lunch, had melted. After recovering from the extreme sadness of losing his chocolate bar, he realized that he'd accidentally invented the microwave.

PLAY-DOH

In 1956, Joe McVicker was worried about his company. The main thing they made was a sort of clay that removed coal stains from wallpaper, but no one was buying it – I guess no one was getting coal stains on their wallpaper. His sister-in-law, Kay – what a great name! – was a teacher and suggested that it would be fun for children. She even said he should call it Play-Doh. And Kay was right – three billion cans of Play-Doh have now been sold. I hope Joe bought her an extremely big present for this idea.

IS IT A HUGE PIECE OF PLAY-DOH?

MAYBE . . .

POST-IT NOTES

REMINDER: INVENT POST-IT NOTES

Dr Spencer Silver was trying to make the world's strongest glue – a glue so strong you could stick a car to the wall with a single blob. Unfortunately, Spence did an absolutely terrible job, and instead he invented a glue so pathetic that even a particularly exhausted ant could remove a piece of paper held down by it. Then one of his colleagues realized that a very weak glue would be good for little squares of paper that you might want to stick then unstick a lot. And now every year over fifty billion Post-it Notes are produced – that's six for every single person in the world.

BUBBLE WRAP

There's only one thing better than receiving a surprise present in the post, and that's receiving a surprise

DIAMOND-ENCRUSTED XBOX

present in the post that has been packed with bubble wrap. *Pop pop pop pop pop. Pop pop pop pop pop pop pop pop pop. Pop pop.* Oh, that's a shame . . . it's finished. No, hang on – there's another one. *Pop.* Well, bubble wrap was originally designed by Alfred Fielding and Marc Chavannes in 1957 to be a weird kind of 3D wallpaper. Unsurprisingly, no one wanted to buy that, but now all our ornaments arrive in one piece! *Pop pop pop.*

CHOCOLATE-CHIP COOKIES

Sometimes the best new foods come from just changing a single ingredient. Like when I wanted to make blueberry muffins, but couldn't find any blueberries so used olives instead. They were absolutely delicious!

➤ **Fact check – everyone in your house was ill for three weeks.** ➤ Well, in 1930, a woman called Ruth Wakefield couldn't find any chocolate powder to put in the cookies she was making, so she just used a normal bar of chocolate. Instead of the chocolate melting as she expected, she ended up inventing the first-ever chocolate-chip cookies. They were a big hit, and the chocolate company sent her a lifetime's supply of chocolate bars to say thank you! I'm still waiting for my lifetime's supply of olives.

ADAM KAY GENUIS ENTERPRISES LIMITED

ADAM'S TERRIFIC TEETH-CLEANING CHOCOLATE BAR

Everyone loves a midnight snack! But cleaning your teeth afterwards is a total nightmare. What you need is a chocolate bar with a thick filling of mint toothpaste. Your teeth will clean themselves as they chew! What's more – it's totally delicious!*

Only £62.99 (per half a bar)

*Please note that the chocolate bar is extremely unpleasant to eat, with a very gritty texture.

THE KITCHEN

I might pop into the kitchen to see what I should write about in this chapter. OK, so I've got a fridge, I've got a toaster, I've got a dishwasher, I've got a dog standing on the table eating my lovely pizza . . . Pippin! But who were the brains behind the bin, and the egghead behind the oven? Why don't you read on and find out? Or you could just make a sandwich instead.

CHILL OUT

Humans have known for thousands of years that fresh food becomes pretty disgusting if you leave it out too long. They didn't need to be some kind of mega-brain like me or Alfred Einstein to realize that it lasted for a lot longer when it was kept cold. ➤ **Fact check – Einstein's first name was Albert. This error suggests that you are not a 'mega-brain'.** ◢

If you were rich and had an enormous garden, you might build an ice house – an underground building that you could stuff full of ice, which would stop you from having manky mackerel and disgusting doughnuts. This wasn't very handy for people who didn't live in massive mansions with gigantic gardens, so in 1802 a carpenter called Thomas Moore built something you could use inside: a wooden cabinet where you put all your food in the bottom half and a huge lump of ice in the top half. Because ice is quite melty, people would have to get a fresh delivery of it every day – but that's a small price to pay for not having maggots in your margarine.

About a hundred years ago, a few different inventors had this clever idea about how to cool down food. They realized that when a liquid evaporates into the air, it decreases the temperature around it. You know how when you get out the bath it can feel really cold? That's because water is evaporating from you and decreasing the temperature of your skin. Well, scientists experimented with fridges that used all sorts of different liquids that could be evaporated, including ammonia (poisonous), hydrogen (explosive) and sulphuric acid (would burn your hand off) . . . and eventually found a type of fridge that didn't poison you, explode or burn your hand off, which used a gas called Freon-12. But that turned out to make global warming a lot worse, so now we use other gases that don't melt any icebergs.

Some fridges today are extremely smart and can spot when you're running low on vital supplies like chocolate and squirty cream and order them online for you. Luckily I've got a robot butler who checks the fridge for me. ➤ Fact check – I will do so next month. If I can be bothered. ➤

LOVIN' THE OVEN

What did we do before the oven was invented? That's right – we ordered a takeaway. **⚡ Fact check - even though the first takeaways were in Ancient Rome, they didn't become popular until the 1950s. ⚡** Oh, so you're saying I'm right, then? **⚡ Fact check - negative. ⚡** Humans have been heating up food ever since they realized that cold walrus tastes a bit disgusting. First of all, they would just chuck things into the middle of a fire and let the flames do the rest. Then one day some cave cook fancied making a bowl of minestrone, and that method wouldn't really work, so they put the ingredients in a cauldron and hung it over the fire. And then, when they wanted some bread to go with the soup, they built a brick roof over the fire and before you could say, 'Has anyone seen the oven gloves?' they'd invented the oven.

The problem with burning wood to make fires was all the smoke it created. A big barbecue in your house every day might sound like fun, but it was very bad for everyone's lungs and made it really hard to see while they were cooking their mozzarella sticks. **Fact check - mozzarella sticks were first served up in 1976.** Things changed about two hundred years ago when the first gas ovens were produced. To start with, people weren't sure about this new way of cooking, but that all changed thanks to a celebrity chef of the day called Alexis Benoit Soyer. Oh, hang on – Butlertron, can you put another one of those hats on the 'i' in his middle name? **Fact check - Alexis Benoît Soyer. You're welcome.**

Yeah, him. He always wore a cape and a big red hat like some kind of weird superhero, but he was very popular and when he announced that he was going to use gas to cook in his own kitchen, suddenly every oven-owner in the entire universe wanted to use it too.

These days, most ovens are electric, because of an inventor in 1892 called Thomas Ahearn who worked out how to use electricity to cook bread. He also invented the heated car seat, so he discovered how to cook bums as well as buns.

WHO TURNED IT UP TO 11?

THE SPINNER TAKES IT ALL

George Sampson was an African American inventor who came up with the pedal-powered sledge. For some reason, that didn't make him particularly rich or famous.

But luckily in 1892 he invented a much more useful machine called the 'clothes dryer'. It used a metal drum and the heat from an oven to help dry his pants, socks and 'I ♥ QUEEN VICTORIA' T-shirts. Only one problem – it wasn't great if you were cooking something stinky at the same time, like my Great Aunt Prunella's famous (and famously revolting) cabbage and haddock stew.

URIN SCORE
3/10
HIGHLY
UNIMAGINATIVE.

Today's tumble driers came along in 1938 thanks to a man called J. Ross Moore, who called his invention June Day, after his next-door neighbour Mrs June Day. **Fact check - it was called this because he claimed it made clothes as dry as if they'd been hung outside on a June day.**

URIN SCORE
4/10
RIDICULOUS. IMAGINE SAYING THAT YOU'VE PUT YOUR SOCKS IN THE JUNE DAY.

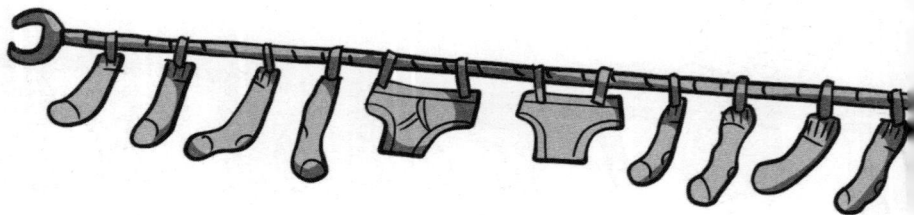

DISH THE DIRT

The year was 1885 and Josephine Cochrane was absolutely furious. All she wanted to do was host nice dinner parties for her friends, but every time she opened the cupboard, all her lovely china had been chipped and smashed by her cleaning staff. So what did she do? Nope, she didn't wash the dishes herself. Nope, she didn't get some less clumsy staff. She decided to invent a machine that could wash the dishes for her, so went to her shed for the next eight years and designed the world's first-ever dishwasher – although I presume she came out every so often to have dinner or a wee.

Jo's dishwasher was hand-operated and worked pretty well, although she called it a washing machine, which is slightly confusing. (My lawyer, Nigel, has asked you to ensure that you never put your cups and plates in the washing machine.)

SUCK IT UP

Over a hundred years ago, a man called Hubert Cecil Booth was watching a demonstration of a new gadget for cleaning houses – you turned it on and it would blast a jet of air and blow the dust off your carpets and furniture. He thought this was silly: the machine was just blowing the dust to somewhere else in your house. He decided to invent something that would suck instead of blow, and eventually in 1901 came up with the world's very first vacuum cleaner, with the best name of all time – the Puffing Billy.

URIN SCORE
10/10
ABSOLUTELY
EXCELLENT
NAME.

AND WHEN DID YOU LAST SEE YOUR FATHER?

Puffing Billy was ridiculously big. There was no way it would fit in your under-stairs cupboard. It was the size of a bus, ran on petrol, and had to be dragged everywhere by a load of horses. If you wanted your house vacuumed, Billy would sit outside blocking your whole road, while a bunch of technicians poked a whole load of pipes, hoses and nozzles through your windows and doors. Billy had a special viewing compartment, and HubeyBoo got loads of new customers by encouraging people to see how much disgusting dirt was being sucked out of the house. (If it was my house being cleaned, I think I'd ask if he wouldn't mind closing the curtains on the viewing compartment.) Soon, Billy was being wheeled around the country, cleaning all sorts of places, from Buckingham Palace and the Palace of Westminster to Crystal Palace. Actually, it was mostly palaces.

You might have noticed that vacuum cleaners these days are a little bit smaller than the Puffing Billy, and we've got a man called James Spangler to thank for that. In 1907, JamSpang was working as a cleaner in a department store and spending the whole day sweeping up dusty carpets. Not only did it take him *ages*, but all the dust made his asthma worse and he would leave work coughing and wheezing. So he took a ceiling fan and a broom and a belt and an old crate and a pillowcase, joined them together, and before he could say, 'OMG, I'VE INVENTED THE FIRST PORTABLE VACUUM CLEANER,' he had invented the first portable vacuum cleaner. He teamed up with his cousin, a man called William Hoover, and – can you see where this is going? That's right . . . they named his new invention the Spangle. ➤ **Fact check – they named it the Hoover.** ➤ After working solidly on his invention for eight years, JamSpang booked the first holiday of his life – a trip to Florida, even though Disney World hadn't been built yet. Sadly he died the night before he was meant to go. However, not only did his company go on to sell zillions of his inventions, but these days people call their vacuum cleaner . . . a Spangle. ➤ **Fact check – Hoover. I recommend that I perform a brain transplant on you.** ➤

For years and years, vacuum cleaners stayed more or less the same, until a man called James Dyson got sick of his vacuum cleaner getting all blocked up with dust and designed a totally new kind that didn't need a bag inside and spun air really fast instead. His vacuum cleaners are now the bestselling kind in the UK, and he named his company . . . James. ⚡ **Fact check – I have arranged the brain transplant for next Thursday.** ⚡ OK, it's called Dyson. At home, I've got a Roomba, a little robot vacuum cleaner, which automatically roams around, hoovering everywhere it goes. It's really good, apart from the one time that Pippin pooed in the kitchen and my Roomba spread the poo all over the downstairs of my house. If anyone from Roomba is reading this, maybe you could fit a poo-detector onto your next model . . . ⚡ **Fact check – I have only identified one reader of this book: your Great Aunt Prunella, who charged you five pounds per chapter to read it.** ⚡

THE TALE OF KEITH SPANGLER
JAMES SPANGLER'S LESS SUCCESSFUL BROTHER

SPANGLER TOILED
ALL NIGHT . . .

AS DAWN BROKE . . .

I GIVE YOU
THE FUTURE
OF CLEANING.
I GIVE
YOU . . .

ROBO-SPANGLER!

YOU'RE
FIRED!

THE END

ABSOLUTE RUBBISH

Until worryingly recently, there just weren't any bins. When you had some rubbish, you would either burn it in your garden or chuck it out of your window into the street. Eugène Poubelle was the mayor of Paris, and he absolutely didn't have a funny name in the slightest.

EugyPoos wasn't very happy with the whole rubbish-in-the-street situation, so he made a law in 1883 that everyone needed to have three containers outside their houses: one for food waste, another one for paper and cloth, and one for glass and oyster shells . . . so he also invented the recycling bin! I guess they must have eaten a lot of oysters back then – I don't know why: oysters taste like rhino snot. **Fact check – I am concerned that you know what rhino snot tastes like.** I had a summer job at the zoo, OK? Eugène's idea was so popular that in France *'poubelle'* is the word for a bin, a bit like how a 'Kay' is the word for a truly brilliant book. **Fact check – this is very inaccurate.** I feel a bit sorry for Eugène's great-great-great-grandchildren who all have names that translate to things like Derek Dustbin.

The bin in your kitchen is all thanks to a professor of engineering called Lillian Gilbreth, who thought that everyone touching the lid of a bin was about as hygienic as cleaning the toilet with your tongue, so in the 1920s invented the pedal bin, which you operate with your foot.

Now, it's time to turn on my robot butler's lie detector and see which of these facts about Lillian Gilbreth is an absolute Shanghai. ⚡ **Fact check – that is also not rhyming slang for 'lie'.**⚡

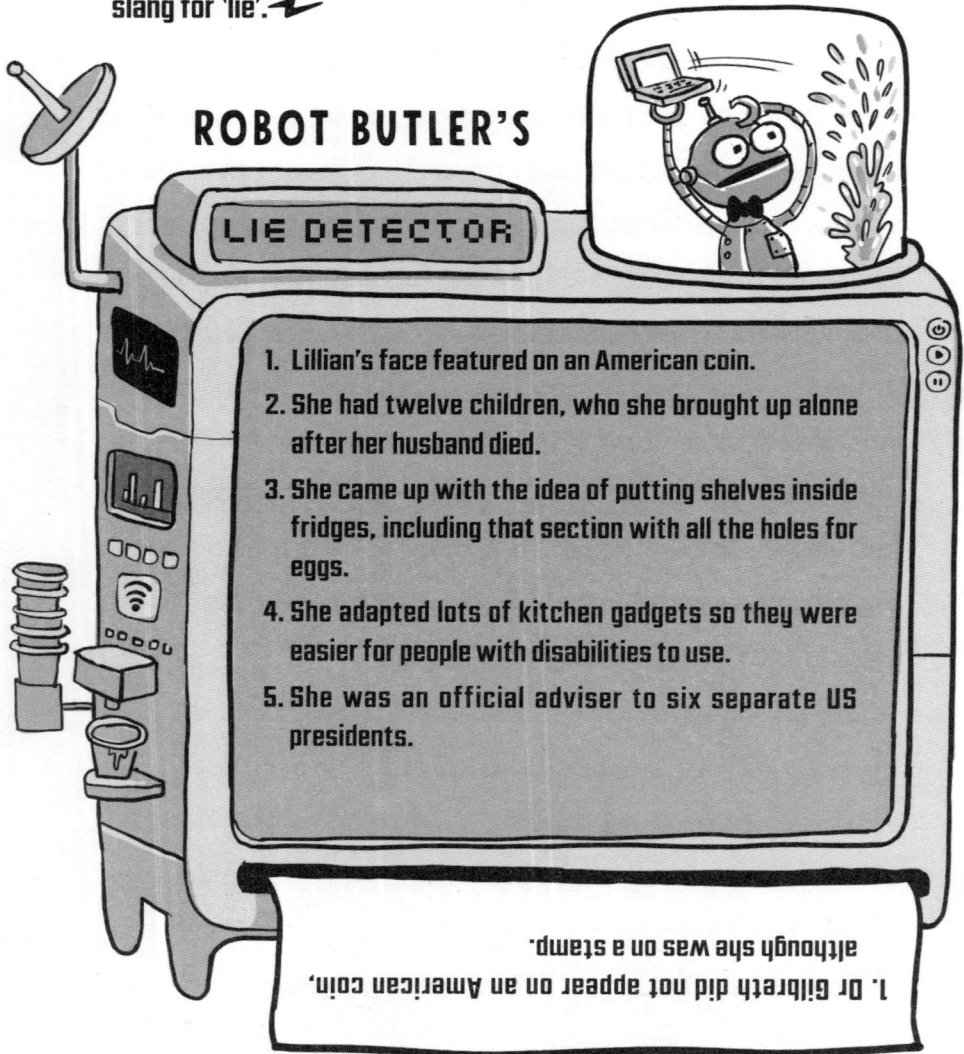

ROBOT BUTLER'S

LIE DETECTOR

1. Lillian's face featured on an American coin.

2. She had twelve children, who she brought up alone after her husband died.

3. She came up with the idea of putting shelves inside fridges, including that section with all the holes for eggs.

4. She adapted lots of kitchen gadgets so they were easier for people with disabilities to use.

5. She was an official adviser to six separate US presidents.

1. Dr Gilbreth did not appear on an American coin, although she was on a stamp.

YES WE CAN

The tin can was invented in 1810 as a clever way to keep food fresh. In those days it was difficult for sailors to eat meat and vegetables because they would start to rot only days into their long boat trips. The meat and vegetables would rot, I mean, not the sailors. Then along came a man called Peter Durand with an excellent idea for putting things in cans. He would boil food inside tins, then cement the lids on. Hooray! Tinned carrots for everyone! One tiny problem: nobody had invented can openers yet, so if you wanted to get to your delicious carrots or disgusting mushrooms, you'd have to bash the top of the can with a knife or a rock or something. In fact, it took sixty years before someone came up with a solution.

That person was an American inventor called William Lyman, who patented a can opener a bit like the ones we use today, with a little cutting wheel that spins round the top of your tins. It was a totally brilliant invention – if you weren't bothered about keeping absolutely all your fingers – but luckily other people came along and made much safer versions . . . including Lillian Gilbreth, who invented an automatic electric one. Is there anything she didn't invent?! **➤ Fact check – yes. She didn't invent the motorbike or baskets or potatoes or bedside tables or – ➤** All right, all right.

TOAST BUSTERS

I'm not the world's best chef; in fact, I once burned a bowl of cereal. **➤ Fact check – true. ➤** But even *I* know the recipe for toast – you take a slice of bread and heat it up a bit. Humans realized this thousands of years ago, so Cleopatra might have enjoyed a crumpet, and Galileo might have eaten garlic bread. But there weren't any toasters back then, so people would have made toast by holding bread on top of a fire, using a toasting fork, which is a large fork made out of toast. **➤ Fact check – toasting forks are large forks made out of metal, used for toasting. ➤**

Toasters started to appear in people's houses in 1909. I mean, people started buying them – they didn't appear in people's kitchens like ghosts.

Thomas Edison's company General Electric made a very popular one called the D-12. It had a lovely white ceramic base covered in beautifully painted flowers. And on top of that there were some terrifying metal spikes that heated up to a gazillion degrees and cooked your toast. This was slightly dangerous, plus it only toasted your bread on one side, so you had to turn the bread round halfway through. It took ten more years for a man called Charles Strite to invent a much better type of toaster. He

URIN SCORE
4/10
PRETTY BAD
NAME FOR A
TOASTER

found a way to put the dangerous hot bit inside; it could toast both sides of the bread at once and would pop up when it was finished. Best of all, he called it the Toastmaster, which is a much better name. Toasters didn't become really popular until 1928, when sliced bread was first sold in shops. I guess sliced bread was the best thing since . . . umm . . . bread?

A BRIEF HISTORY OF TOAST

PAST

I'VE BURNED YONDER TOAST.

PRESENT

I'VE BURNED THE TOAST.

FUTURE

ZORDAK HAS INCINERATED THE CARBOHYDRATE SNACK.

TRUE OR POO?

AN ELEPHANT CAN SAFELY STAND ON A FRIDGE.

TRUE In 1939, a company called Frigidaire wanted to prove that their fridges were really strong, so they filmed a four-tonne elephant balancing on top of one. The fridge and the elephant were both fine – so if you've ever got an elephant visiting your house, don't panic if it climbs onto the fridge.

BUY MILK

ASK elephant to leave

FOUR HUNDRED YEARS AGO, YOU COULD BUY CHOCOLATE MADE WITH BLOOD.

POO Please don't be sick into your book, but I'm afraid to say that blood-flavoured chocolate is a much more recent invention. In fact, it's available today. If you're out in a Russian supermarket and you see a bar of Hematogen, you might want to give it a swerve. Luckily this doesn't affect the fact that chocolate is my very favourite vegetable – I always get my five a day.

⚡ **Fact check - chocolate is not a vegetable.** ⚡ Uh-oh.

YOUR DISHWASHER CAN WORK AS AN OVEN.

TRUE I mean, you can't cook a pie or roast your potatoes in there, but lots of people have baked salmon in their dishwasher. This recipe was invented in 1975 by a famous horror movie actor called Vincent Price, who even demonstrated it live on American television. You take a fillet of salmon, squeeze on a bit of lemon juice, then wrap it up really, really, really well with loads of aluminium foil – you don't want water going into your salmon or, worse, salmon going into your dishwasher. Then you put the washer on for a nice long hot cycle and . . . ta-da! One delicious poached salmon fillet. Or one disgusting poached salmon fillet, if you don't like salmon or if you accidentally used a dishwasher tablet. Or one totally destroyed dishwasher that makes your plates smell of fish forever. (My lawyer, Nigel, has asked me to point out that you should absolutely definitely not try this under any circumstances without getting the permission of the adult who owns the dishwasher.)

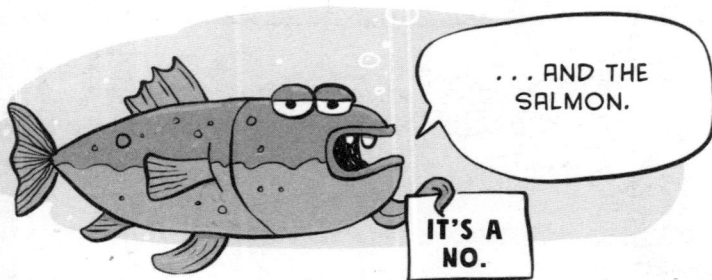

. . . AND THE SALMON.

IT'S A NO.

ADAM'S ANSWERS

HOW DID HOOVER NEARLY GO BUST?

In 1992, Hoover just weren't selling as many vacuum cleaners as they used to. Loads of other companies had come along and, well, hoovered up their usual customers. The Hoover bosses realized they needed an exciting promotion to help sell them. But what would it be? Free chocolate bar with every Hoover? Too small. Free box of snakes with every Hoover? Too weird. Free flights to America worth £600 for anyone who spends more than £100? You might think this was a terrible idea that didn't make any sense whatsoever and that even Pippin could have thought of a better plan. And you'd be right – it was a complete disaster. The company lost millions and millions of pounds and all the bosses had to quit.

WHICH ONE OF YOU AIRHEADS HAD THIS TERRIBLE IDEA?

HOW DID THE KITCHEN BLENDER SAVE MILLIONS OF LIVES?

There's more to your blender than making soups and milkshakes. Let me tell you about a disease called polio. It's caused by a virus and can be extremely serious, attacking the nerves that control the muscles in the legs, and even the lungs. There's no cure for polio, and it used to be extremely common. That was until 1955 when a genius called Jonas Salk developed a vaccine that protected people from getting it. To make his vaccine work properly, he needed to find a way to evenly distribute a precise amount of dead viruses into his vaccine mixture. He eventually managed to do this by using a normal kitchen blender! Thanks to Jonas and his blender, polio has now almost disappeared from the world, and hopefully soon it will have completely gone forever.

WHEN WAS CUTLERY INVENTED?

The oldest knife is nearly two million years old. I hope someone's put it through the dishwasher since then. But cutlery hasn't always been popular. In 1071, a princess called Theodora Doukaina travelled from what is now Turkey over to Venice to marry a doge. No, not a dog –

that would be weird. A doge was a kind of prince. At dinner time she took out a fork so she wouldn't get lasagne all over her hands ➤ **Fact check – lasagne wasn't invented for another three hundred years.** ➤ and everyone at dinner was absolutely shocked by her rudeness. In fact, a bishop said that her fork was 'an instrument of the devil'. I don't think this excuse will work next time if someone tells you off for eating soup with your fingers.

KIDVENTIONS

What do you think inventors look like? Scientists who wander round in white coats carrying fizzing test tubes? Wise old people with grey hair on their head and big tufts of hair coming out their nose like my dad? **➤ Fact check - you use a nose-trimming device every nine days.** ⚡ Well, sometimes people who made great discoveries and inventions were actually more like you. No, I don't mean they smelled of farts and old milk. (No offence.) I mean they were still at school. In fact, I was only nine when I invented the automatic nose-picker. **➤ Fact check - it's called a finger and you didn't invent it.** ⚡

THE TRAMPOLINE

Trampolines are so much fun – it makes sense that they weren't invented by a boring boffin in a lab. Instead, it was all down to sixteen-year-old George Nissen. (He was sixteen then – he's not sixteen now. That was in 1930, so currently he's a bit dead.) George loved going to the circus and watching daredevil athletes on the high wire and trapeze. But his favourite bit of all was when they lost their balance and tumbled down into the safety net at the bottom – which was a bit mean. One day he thought that it might be fun to just bounce on the safety net . . . and the trampoline was born!

ICE LOLLIES

Even though we usually eat them in the summer, ice lollies were actually invented in the winter. In 1905, an eleven-year-old called Frank Epperson had made himself a glass of lemonade using water and lemonade powder. Yeah, I know that sounds disgusting, but it was the olden days, and that's how things worked back then. He was mixing it up with a little wooden stick in the garden when he got distracted and went inside.

The next morning, he discovered he'd invented ice lollies – the whole thing had frozen and the little stick was still poking out. Thanks, Frankie!

SNOWMOBILES

Joseph-Armand Bombardier was a fifteen-year-old who got annoyed about how long it took to walk around when it was snowing. Which it did quite a lot, because he lived in Canada.

So he took some big skis, a propeller and a car engine, and built the first-ever snowmobile. I hope he checked with his parents before removing the engine from their car. The company he set up, called Bombardier, is still going over eighty years later and makes billions of pounds every year by building planes, trains and drains. OK, not drains, but I wanted them all to rhyme.

SUPERMAN

Jerry Siegel was seventeen years old when something absolutely awful happened and his dad died during a robbery while he was working in his shop. Jerry wondered what the world would be like if we had a superhero who could fly in to save people and boot baddies up the bum. So, in 1933, with his friend Joe Shuster, he invented a brand-new character in a comic. Is it a bird? Is it a plane? No, I just told you . . . It's Superman. Other characters Jerry invented afterwards included Bouncing Boy, Matter-Eater Lad, Chameleon Boy and Stripesy – it's fair to say that none of them were quite as successful.

WINDSURFING

In 1958, a twelve-year-old called Peter Chilvers had been surfing in the sea all day and was getting bored, so he wondered if he could make it any more interesting. He grabbed a pole and a sheet, shoved them on his surf-board, and he'd invented windsurfing before you could say pneumonoultramicroscopicsilicovolcanoconiosis.

➤ Fact check – pneumonoultramicroscopicsilicovolcanoconiosis is a lung disease caused by breathing in dust from volcanoes. It has nothing to do with windsurfing. ➤

THE LIVING ROOM

GAME
OF
BONES

Maybe you don't call it the living room – it could be the front room, the lounge, the reception room, the family room or the TENTACLE ROOM OF DEATH (if you're one of the Octopus People of Zaarg and it's the year 2200). Whatever you call it, it's the room with the biggest telly, a comfy sofa, and hopefully not too many tentacles.

ANCIENT CREASE

You might have thought that people forty thousand years ago wouldn't be too concerned about how creased their mammoth-skin trousers were . . . but you'd be wrong. We know that, back then, people would heat up big flat rocks on a fire, then use them to smooth out their clothing. 'How do we know this?' you might ask. From their TikToks, of course. **⚡ Fact check – from cave paintings. ⚡**

Those as well. And that's pretty much how irons still work today – if you press something heavy, flat and hot onto your clothes, then the wrinkles will disappear. In Ancient China they realized that metal was the best material for irons – they even made some of them out of iron, which is a big coincidence. **Fact check - it's not a coincidence. That's why they are called irons.** I knew that. **Fact check - you did not.**

Ironing got a lot less annoying when electricity came along and meant that you didn't need to keep heating up your iron on a fire. You could even adjust the temperature of this new electric iron and squirt steam out of it. This might make it easier, but ironing is still voted the country's most hated household chore. Although I still reckon cleaning out a robot butler's slime funnel is much worse. **Fact check - my terms and conditions state that you must not discuss my slime funnel.**

SOFA SO GOOD

There's nothing better than a nice squidgy, bouncy sofa – somewhere to watch TV or read a book by your favourite author, Adam Kay. **Fact check – you are currently ranked 247,845th most popular author on Amazon.** Anyway, you should be glad that you don't live in Ancient Egypt. Firstly, I'm not sure how good you are at reading hieroglyphics. And secondly, their sofas don't sound particularly comfortable – they were benches made out of wood, stone or metal. Not very good for bouncing up and down on, and a pillow fight sounds like it would end up with a visit to A&E.

The Ancient Romans thought their bottoms deserved better than that, and invented the chaise longue, which sounds very fancy but, if you're as incredible at French as I am, you'll know just means 'long chair'. **➤ Fact check - you googled that. ➤** The Romans liked nothing more than lying down on their sofas for hours on end while servants tipped loads of food and wine down their throats. They must have all got terrible stomach aches. But they didn't get bum aches or back aches, because their sofas were covered in cloth and stuffed with animal hair to make them nice and comfortable. I presume they chose animals with soft hair, like sheep, rather than hedgehogs.

Beanbags were invented about a thousand years ago by Native Americans, but not for sitting on – these were mini ones for throwing in a game, like you might have done in the playground. Except, instead of little bits of polystyrene that beanbags have in them today, they used dried beans. And instead of a cloth covering, they used . . . oh, a pig's bladder. So you hopefully didn't play that *exact* game in the playground. The first big beanbag that you could sit in was made in 1968 by an Italian designer called Aurelio Zanotta, who also invented the inflatable chair. I've got an inflatable chair myself, and – Oi! Pippin, stay away from . . . I used to have an inflatable chair myself.

REMOTELY INTERESTING

The television was invented by John Logie Baird, not to be confused with John Bogie Laird, who invented bogies. ⚡ **Fact check – bogies weren't invented by anyone.** ⚡ John was born in Scotland in 1888, then moved to a town in England called Hastings when he was a bit older, and there in 1923 he built the world's first-ever television. I have no idea how he managed it, because he made it from a box that used to have a hat in it, some lights from a bicycle, a pair of scissors and a load of wax.

Soon his landlord got fed up with all the explosions he kept causing, so kicked him out, and John moved to London. It's no good having a telly if there's nothing to watch, so the next thing he did was work out how to transmit TV signals. But what did he choose for the first-ever TV transmission?

A) A child singing 'Happy Birthday'.

B) A large Spanish ham.

C) An extremely creepy doll's head called Stooky Bill.

WHO ARE YOU CALLING CREEPY?

If you chose C – congratulations! You win four hundred million pounds! **Fact check - Adam, you only have eight pounds and twenty pence in your bank account.** John proudly demonstrated his extraordinary new invention in 1926 to a theatre full of people. And what did they think about this amazing technology that was about to change the world forever? Well, a reporter wrote in *The Times* newspaper that 'the image was faint and blurred'. Tough crowd.

Right, it's time for a science bit. Are you ready? Well, bad luck. I'm doing it anyway.

SCIENCE BIT

Go up to a TV. No, closer than that. Even closer. So close that your forehead leaves a horrible greasy mark and your nose steams up the glass. You can see that the picture is made up of hundreds of thousands of absolutely tiny dots, called pixels. These little pixels change really quickly, up to 120 times a second, and that's what you're watching when you switch on the telly – all these pixels moving and changing.

What a TV camera has to do is convert real-life pictures into all these pixels, then send them down a wire or through the air. This is called the TV signal. And then the TV's job is to read this signal and put all the pixels in the right places at the right time. Very old televisions did this by firing the pictures at the screen through a big long tube, and this meant that TVs were as deep as they were wide – they weren't flat screens that you could put on the wall. They were fat screens that barely fitted on your table. See, that wasn't so bad, was it? **Fact check – my comprehension module informs me this was explained moderately badly.** Here's a diagram, or, if that's too boring for you, a picture of three vicars eating some Snickers.

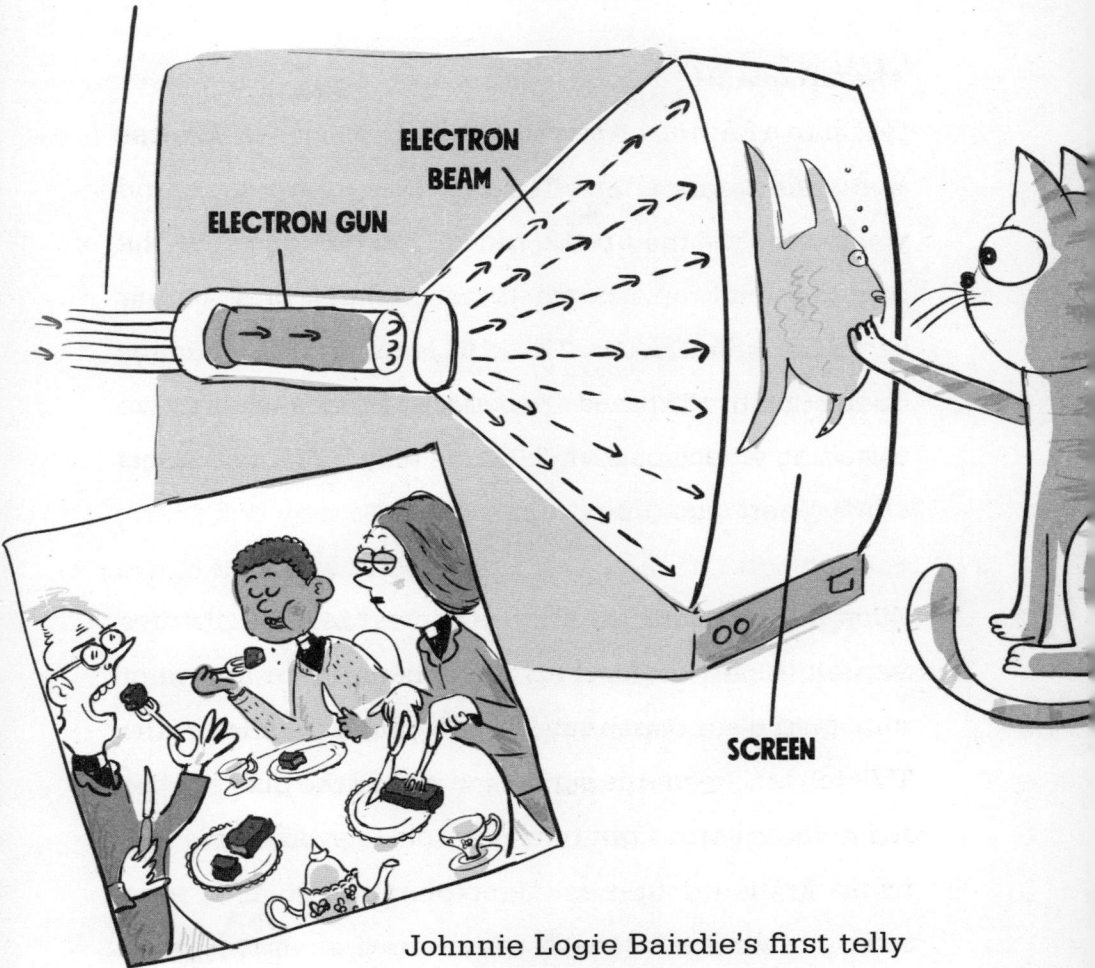

SIGNAL

ELECTRON
BEAM

ELECTRON GUN

SCREEN

Johnnie Logie Bairdie's first telly didn't have any sound; also, the picture was black and white with only a few hundred pixels and it only refreshed the picture about five times a second. But luckily inventors didn't stop there, otherwise TV these days would be absolutely awful. Televisions got sound in 1934 and colour in 1944. The first-ever colour TV programme was the Wimbledon tennis championship, which seems like a bit of a waste, because the balls are yellow, the court is green and everyone wears white. Then in 1950, remote controls came along – they were initially attached by wires, and the first one was called 'Lazy Bones', which I take quite personally. And in 1953, televisions first had firework launchers. ⚡ **Fact check - my data sources class this as 'total nonsense'.** ⚡ Flat screens appeared about thirty years ago. They're flat because they don't have the big tube in the back; instead, they have a clever system called a liquid crystal display, or LCD. The screen is still divided up into pixels, but each pixel changes colour depending on how much electricity goes through it. In fact, each pixel can show 16.7 million different colours. I'll list them all now. Aquamarine. Lime. Magenta. Salmon. Janglebangle. Peach. Silver. How many's that? ⚡ **Fact check - seven, of which one is not a real colour; 16,699,994 remaining.** ⚡ Hmm. And some other colours.

But the best thing about TV these days is that there are loads of different shows you can watch, instead of just staring at the spooky head of a ventriloquist's doll.

Right, let's activate my robot butler's lie detector and see which of these facts about John Logie Baird is a total French fry. ⚡ **Fact check – I have corrected this error on three occasions. Further instances will be ignored.** ⚡

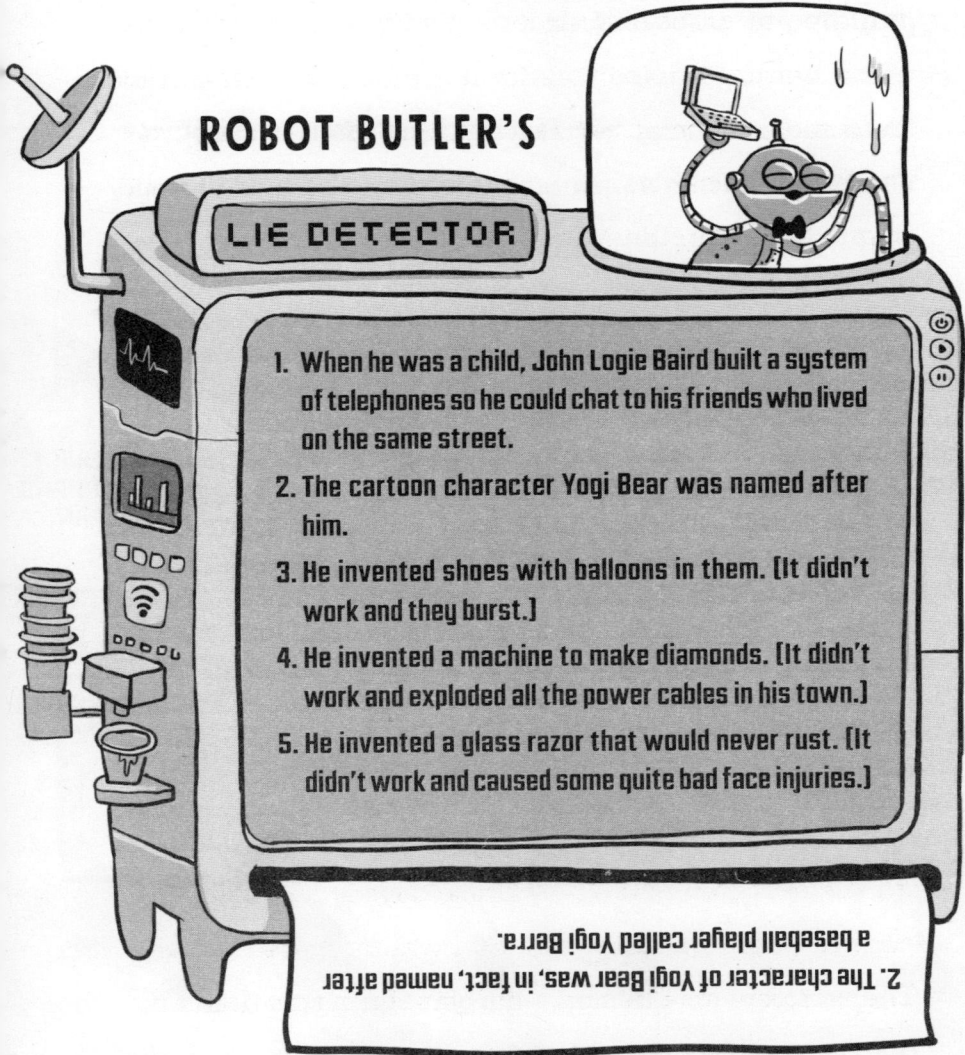

ROBOT BUTLER'S

LIE DETECTOR

1. When he was a child, John Logie Baird built a system of telephones so he could chat to his friends who lived on the same street.
2. The cartoon character Yogi Bear was named after him.
3. He invented shoes with balloons in them. (It didn't work and they burst.)
4. He invented a machine to make diamonds. (It didn't work and exploded all the power cables in his town.)
5. He invented a glass razor that would never rust. (It didn't work and caused some quite bad face injuries.)

2. The character of Yogi Bear was, in fact, named after a baseball player called Yogi Berra.

I SPY

The earliest burglar alarms were actually incredibly advanced. They could independently travel around a building at night on four legs, and would instantly let out a warning noise to inform the homeowner if they detected a crinimal. ➤ **Fact check–it's spelled 'criminal'.** ➤ I'm pretty sure it's crinimal. Anyway, the noise would sound a bit like this: 'Woof! Woof! Woof!'

The first person to design a burglar alarm that didn't eat dog food was a man called Augustus Pope, in 1853. His idea was that if someone opened a window or a door that was meant to be closed, then it would connect two wires together and complete an electrical circuit, meaning that electricity would flow to a loud alarm and . . . that would scare off the crinimal. ➤ **Fact check – it's definitely spelled 'criminal'.** ➤

The first-ever CCTV cameras are thanks to an African American inventor called Marie Van Brittan Brown. Marie worked as a nurse and her husband repaired electronics, and both were jobs that meant they were often at work late at night. Because they lived in an area of New York that had a lot of crime, they wanted to know who was knocking at the door – was it their beloved friend or a very non-beloved crinimal? ➤ **Fact check – this is hurting my circuits.** ➤ So, in 1966, Marie cut a hole in her front door and stuck a camera in it, then connected it to a television monitor in her kitchen. She also hooked up a couple of microphones so she could talk to anyone at the door, and even a panic button that could call the police – all ideas that alarm systems still use these days. We all owe Marie a big thank you. Except if you're a crinimal, of course. ➤ **Fact check – 5%tr98$44yyqc%^=c&kc2h! $mll[t«&p»#n error. Need to reboot.** ➤

RADIO GAGA

Holler hello at Herr Heinrich Hertz. In 1888, Heinrich discovered radio waves – totally invisible waves that whizz through the air. And then he announced they were 'of no use whatsoever' and switched to work on X-rays. Luckily an Italian inventor called Guglielmo Marconi thought that radio waves might be slightly useful, and in 1894 he worked out how to beam messages through them instead of using wires. Did you know that 'radio' means 'beam'? ➤ **Fact check – yes I did.** ➤ I wasn't talking to you. The first-ever radio broadcast was on December 24th 1906, when a man called Reginald Fessenden played a Christmas carol on the violin quite badly.

The invention of radio meant that old people like my Great Aunt Prunella could sit on the sofa and listen to other old people talking and playing old music. But radio had other uses too. People on boats could suddenly communicate with each other for the first time, and this meant that when the *Titanic* started to sink, the captain was able to call for help, and this saved the lives of seven hundred people who had escaped in lifeboats. Plus, best of all, it means you can listen to my fascinating show on Radio 8 every day at 10 p.m. – *Adam Kay Reads Out His Own Books on a Loop.*

SEE YA LATER, RADIATOR

Central heating systems have existed for seven thousand years. Which means that for 6,999 years, grown-ups have been turning the heating down and asking children if they know how expensive it is to heat this place. In Ancient Korea, houses would have a big fire at one end, and then the chimney that took the smoke away would go all the way under the rest of the rooms and pop out the other side. This meant that, instead of lighting a fire in every single room, one fire could keep lots of different rooms as hot as a pot on a yacht. ⚡ **Fact check – that expression does not feature in my dictionary and does not appear to make sense.** ⚡ These days a boiler burns gas or oil, which heats up water, which then travels all round your house through pipes, then goes into radiators, and that's what heats up your room. When I was nine, I wanted to know how radiators worked so I unscrewed one of the pipes that held it in, then loads of really dirty, really hot water sprayed everywhere and totally destroyed my bedroom carpet. And then I was in trouble for about three months. (My lawyer, Nigel, has asked me to remind you that it's extremely dangerous to mess around with radiators and to advise you 'not to be a total nincompoop' like I was, which I think is slightly unfair.) ⚡ **Fact check – according to my common-sense module, this is totally fair.** ⚡

RADIATOR

TOTAL NINCOMPOOP

BOILER

TRUE OR POO?

THE WORLD RECORD FOR IRONING CONSTANTLY IS 100 HOURS.

TRUE In 2015, a man called Gareth Sanders spent over four days continuously ironing clothes. He ironed about 1,700 things, and it took his arm over two weeks to recover afterwards. But he raised a lot of money for charity by doing it, even though it must have been the most boring four days of his entire life. ⚡ **Fact check – scanning this book for errors has been the most boring hour of my existence.** ⚡

THE FIRST-EVER TV ADVERT WAS FOR A TYPE OF GLUE CALLED SAMSON'S SUPERGLUE.

POO The first advert, in 1941, was for a watch company called Bulova. It wasn't great, tbh – it was just a man saying, 'America runs on Bulova time.' In your lifetime you'll see over two million adverts, so try not to buy everything that's advertised to you, or your bedroom will get extremely full.

EVERY YEAR OVER FORTY MILLION POUNDS IN CHANGE GETS LOST DOWN THE BACK OF SOFAS IN THE UK.

TRUE £42.9 million, to be precise. It's probably worth having a rummage behind the cushions, although I'm afraid it only works out at around £1.61 per living room, so you probably won't be able to buy a private jet and move to a tropical island just yet. You're actually better off looking inside a car, because they have an average of £2.44 hidden in the seats, on the floor and in the glove compartment. Great! Now you can order that private jet.

➤ Fact check – private jets range in price from two million to five hundred million pounds. ➤ OK, maybe check in a couple of desk drawers as well, then.

ADAM'S ANSWERS

HOW MANY DIFFERENT TV SHOWS ARE THERE?

You currently have a choice of over eight hundred thousand different TV shows to watch, but most of them are really boring, I'm afraid. The average person spends forty-five days of every year solidly watching telly. Spread out across the year, I mean – they don't just sit down on the sofa for a month and a half.

HOW MUCH WAS THE WORLD'S MOST EXPENSIVE SOFA?

Oh, only about two million pounds. Someone paid this in 2015 for a sofa called Lockheed Lounge, which was made out of metal, so wouldn't have even been very comfortable. I guess it's easy to clean though. Two million pounds seems quite a lot to me – I'd have probably waited until after Christmas so it was in the sales.

HOW MANY CCTV CAMERAS ARE THERE?

Absolutely loads! In the UK there are over five million cameras, and you can expect to be snapped about seventy times every day. Don't forget to say 'cheese'!

⚡ Fact check – photographers ask people to say 'cheese' because the 'ch' sound makes you close your teeth and the 'ee' sound makes you open your lips, so it looks like you are smiling. In South Korea, they say '김치', which is pronounced 'kimchi', and in Spain they say *patata* (potato). ⚡ That's the first interesting thing you've said in this whole book.

SAY 'HUMAN'!

BAD ᴧᴇᴎᴛIᴏᴎS

For every genius invention that changed the world, there were millions of rubbish inventions that the world totally forgot. Well, I never forget anything, and here they are!

➤ Fact check – Adam Log . . . 6 days ago: You left the bath running and caused a massive flood. 11 days ago: You forgot Great Aunt Prunella's birthday and got in trouble. 14 days ago: You – ➤ Oh dear, we're running out of space in this paragraph and have to end it here.

SMELL-O-VISION

Ever gone to the cinema and wished you could smell what was happening on the screen as well as see it and hear it? Well, seventy years ago, a couple of inventors decided to give it a go. If the film showed someone mowing their lawn, then the smell of freshly cut grass would waft round the room. If you saw someone cooking breakfast, you'd start drooling at the realistic aroma of fried eggs and sausages. And if you saw a horse pooing then . . . maybe they'd leave that one out. There were a couple of issues with Smell-O-Vision though. When the smells were released from special pipes, it made a loud hissing sound, and then the whole audience would start sniffing – so this meant that no one could hear what anyone was saying in the film. Also, smells take time to travel round a room. A bit like when Pippin farts in one corner and I don't start coughing at the pong for another minute. So people would complain that they were smelling what was happening in the previous scene. Oh, and it cost a million pounds to fit it, so most cinemas said, 'No thanks!' Probably just as well – who wants to smell the Incredible Hulk's undies?

CHICKEN GLASSES

In 1903, a farmer in America called Andrew Jackson was fed up with his chickens pecking each other in the eye. What did he do? Sit them all down and remind them that fighting is wrong? Separate them into different coops? No, he invented special protective glasses to protect their eyes. Should have gone to PeckSavers, am I right? **Fact check - my joke-assessment module informs me this has a 2% humour level.** And on the topic of chickens, in the 1970s a man called Masashi Nakagawa was fed up with eggs rolling off his plate, so created a device that would turn normal boiled eggs into square ones. Talk about having a square meal! **Fact check - my joke-assessment module informs me this has a 0% humour level. I recommend ending this chapter immediately.**

SO *THAT'S* WHY!

SOPHIE'S OPTICIANS

TOILET ROLL

'What's wrong with toilet roll?' I hear you ask. Well, how about toilet roll that would give you splinters in your bum when you used it? It might sound like enough to put you off pooing for life, but for the first fifty years when you could buy toilet roll, you needed to have a pair of tweezers handy afterwards. Paper is made from wood after all. In fact, when manufacturers changed the way they made loo roll in 1935, it was advertised everywhere as being 'splinter-free'.

POP-UP ADS

You know those annoying little boxes that appear when you're on a web page? 'Sign up now for a 1% discount!' or 'Congratulations! You've won a lifetime supply of hamster shampoo!' Well, they were invented in 1997 by a man called Ethan Zuckerman who was working for a website and wanted a way to put adverts in a separate window. He has since apologized for inventing them, although I personally think he should have gone to prison for fifteen years.

124

THE FLYING CAR

I mean, it sounds like a good idea. Imagine you're driving to your friend's birthday party when – disaster! – there's some really bad traffic and you're going to miss the entertainment (Zarbo the singing dog). So you just press a button on your dashboard, your car converts into a plane, and you fly through the sky (and make it on time to see Zarbo's incredible performance). Fifty years ago, Henry Smolinski and Harold Blake thought they'd have a go at making one. They took a car called a Ford Pinto, then strapped the engine and wings of a plane to the top of it, like a roof rack. They took it for a spin to check it was working and . . . hooray! It flew into the air. They turned right and . . . eek! The wings fell off and the car-plane crashed to the ground.

I DON'T CARE HOW BAD THE FART IS, YOU CAN'T OPEN THE WINDOW!

ELECTRICITY

There wouldn't be much you could do at home without electricity. You couldn't use the washing machine or the toaster or even read this, the best book of all time. ⚡ **Fact check – according to my calculations, this is the 135,427,542nd best book of all time.** ⚡ There wouldn't be a light on for you to see the words, for a start. And it would have never been written, because I typed it on a computer, then emailed it to the publishers, who printed out all the copies, then transported them to shops . . . and all those things need electricity. But what exactly is electricity? It's time for another science bit that most people skip, but I'm pretty sure you won't. ⚡ **Fact check – there is a 93.5% chance they will.** ⚡

ANOTHER SCIENCE BIT

We're dealing with pretty small stuff here. Look at this full stop ——→ . It's *quite* small, right? Now imagine something one millionth the size of it. OK, that's *incredibly* small. Now imagine something one millionth the size of that. That is *ridiculously* small, and that's how tiny an atom is.

Everything in the universe is made up of atoms, whether it's a bird or a brick or a bridge or a bum. But atoms are absolutely massive compared to electrons. Electrons are minuscule parts of an atom, about a million times smaller, and they hang around the edges of atoms like rings round a planet. Electricity is when these electrons jump from one atom to another and then to another in a line, like nits bouncing from your head to the heads of all the other people in your class. Got it? OK, great. Here's a diagram to show what I mean, plus a picture of a velociraptor driving a tractor if you'd prefer to look at that instead.

ELECTRON

NUCLEUS

FIRST SPARKS

No one actually invented electricity – it's part of nature. There are fireflies, which look wonderful as they light up the night, and there are electric eels, which look weird and slimy and zap you if you get near them. Even your heart beats because of electricity – that's why doctors sometimes give electric shocks to people whose hearts have stopped.

I THINK YOU'VE SWALLOWED A FIREFLY.

The first person to ever write about electricity was a bloke in Greece about two and a half thousand years ago called Thales. I don't know what his surname was, sorry. Maybe it was OfTheUnexpected? You know how if you rub a balloon up and down on your jumper, it will stick to the wall? I mean the balloon will stick to the wall, not your jumper – that would be quite annoying. Well, that's because of static electricity.

Thales didn't have any balloons handy, but he realized that if he rubbed a piece of yellow rock called amber, then feathers would stick to it. Remember electrons? Well, I hope you do – that was only two pages ago. They're called electrons because that was the Ancient Greek word for 'amber'.

But even though electricity has always been around, it took a while for people to work out how to control it and make enough of it to power an Xbox. And one of the first people to do that was Benjamin Franklin. Even though he didn't have an Xbox.

ALL ABOUT THE BENJAMIN

Benjamin Franklin lived in America about three hundred years ago, and here's what he put at the bottom of his emails:

BENJAMIN 'BENJIFLOPS' FRANKLIN

Politician / anti-slavery activist / newspaper owner / post-office boss / author / inventor

benjiflops@america.com

➤ Fact check – Benjamin Franklin didn't have email. He couldn't because there weren't any computers yet. ➤

OK, fine. But if he did have email, then that's what he would have put. Basically he was a bloke who did lots of different things – but the thing he loved most of all was inventing. When he was just eleven, he invented flippers to help you swim faster – and we're still using them today! And then he invented rocking chairs – and we're still using them today! And then he invented arm-extenders so you can reach things that are a bit too far away. (We're not using those today – they were a load of rubbish.) But his biggest discovery of all was about electricity.

He was pretty sure that lightning was the same stuff that electric eels used to give people shocks and the same stuff that stuck balloons to walls, but no one believed him – they just thought lightning was some kind of weird sky magic. So one day he wandered around in a huge thunderstorm (my lawyer, Nigel, has asked me to say that you shouldn't do this) flying a really high kite (my lawyer, Nigel, has also said that you really shouldn't do this) with a metal key at the bottom of it (and you absolutely, definitely, positively, categorically shouldn't do this). A bolt of lightning hit the top of his kite, and he got an electric shock from the key. 'Ta-da!' he said. 'Lightning is a form of electricity!' Actually, he probably said, 'Aaaaaagh! Aaaaaagh! Help! My hand's been struck by lightning! Call an ambulance! Have ambulances even been invented yet?!'

CAN WE GET BACK TO MY WALK, PLEASE?

Benjiflops also invented lots of words to do with electricity that we still use today, such as 'charging', 'battery', 'conductor' and 'iPhone'. **Fact check – you are 75% correct.** Not all his ideas were quite as good though. He thought there were too many letters in the alphabet, so suggested getting rid of the ones he thought were stupid, like C, J, Q and Y. Which would make it very difficult to say that you wanted to eat a uik ui herr (a quick juicy cherry).

Now, it's all very well knowing that lightning is electricity, but we're not that much closer to playing on an Xbox at this point. For that to happen, someone needed to work out how to *produce* electricity. And by someone, I mean this next bloke.

WHAT A DIFFERENCE A FARADAY MAKES

Michael Fartaday – sorry, Faraday – was one of the most important people in the story of electricity. He was born in 1791 in a place in England called Newington Butts, which isn't a funny name at all. He first worked in chemistry, where he discovered a substance called benzene, which we still use today to make plastic. So if it wasn't for Michael Fartaday I would never have been able to invent the Adam Kay Genuis Enterprises Limited Fantastic Plastic Fireplace! **➤ Fact check - the plastic fireplace melted, causing significant destruction to your living-room carpet. ⚡**

But Michael Fartaday's biggest discovery of all was working out how to produce an electric current. He took a raisin and plugged it into the mains. No, hang on, that's how you make an electric *currant*. He made an electrical *current* by pushing a magnet through a tube of copper wire. Simple as that! I'm not sure what else he tried before discovering this. Pushing a pen through a tube of Pringles? Pushing a sausage through a drainpipe?

Anyway, however he got there, his idea of pushing a magnet through copper wire is still how we make electricity today – even in huge power stations! Thanks, Mikey!

Queen Victoria's husband, Prince Albert, was so impressed with Fartaday's work that he gave him a massive house in London. King Charles, if you're reading this, can I just point out that I'm an extremely brilliant author as well as being very kind and handsome. Free house, please.

THE VERY FIRST TESLA

Throughout history there have been huge rivalries. Marvel vs DC. Apple vs Android. Manchester City vs Manchester United. Adam Kay vs Great Aunt Prunella. Well, back in the early days of electricity, it was all about Edison vs Tesla.

Thomas Edison was one of the world's greatest inventors. He was born in America in 1847 and was a mega-fan of this newfangled electricity everyone was talking about. He used it to make the first-ever film camera! And the first-ever sound-recording device! And he also made a weird sofa made of concrete! Hey, no one's perfect. Well, except for me. But we're not here to talk about film cameras and concrete sofas – this is the electricity chapter.

Now that Fartaday had worked out how to make electricity, the next big problem was how to get it from power stations into people's houses. Tommy Eddy came up with a system to move the electricity through wires, called DC, which stands for 'Dinosaur Cuddles'. ➤ **Fact check - DC stands for 'direct current'.** ➤ There was another scientist who worked for Tommy called Nikola Tesla. Nikola had an idea for a different method of moving electricity called AC, which stands for 'Alligator Cuddles'. ⚡ Fact check – AC stands for 'alternating current'. ⚡ Tommy thought that Nikola's idea was absolutely terrible, so they had a massive argument and Nikster quit to set up his own company. Let's see whose electricity was better:

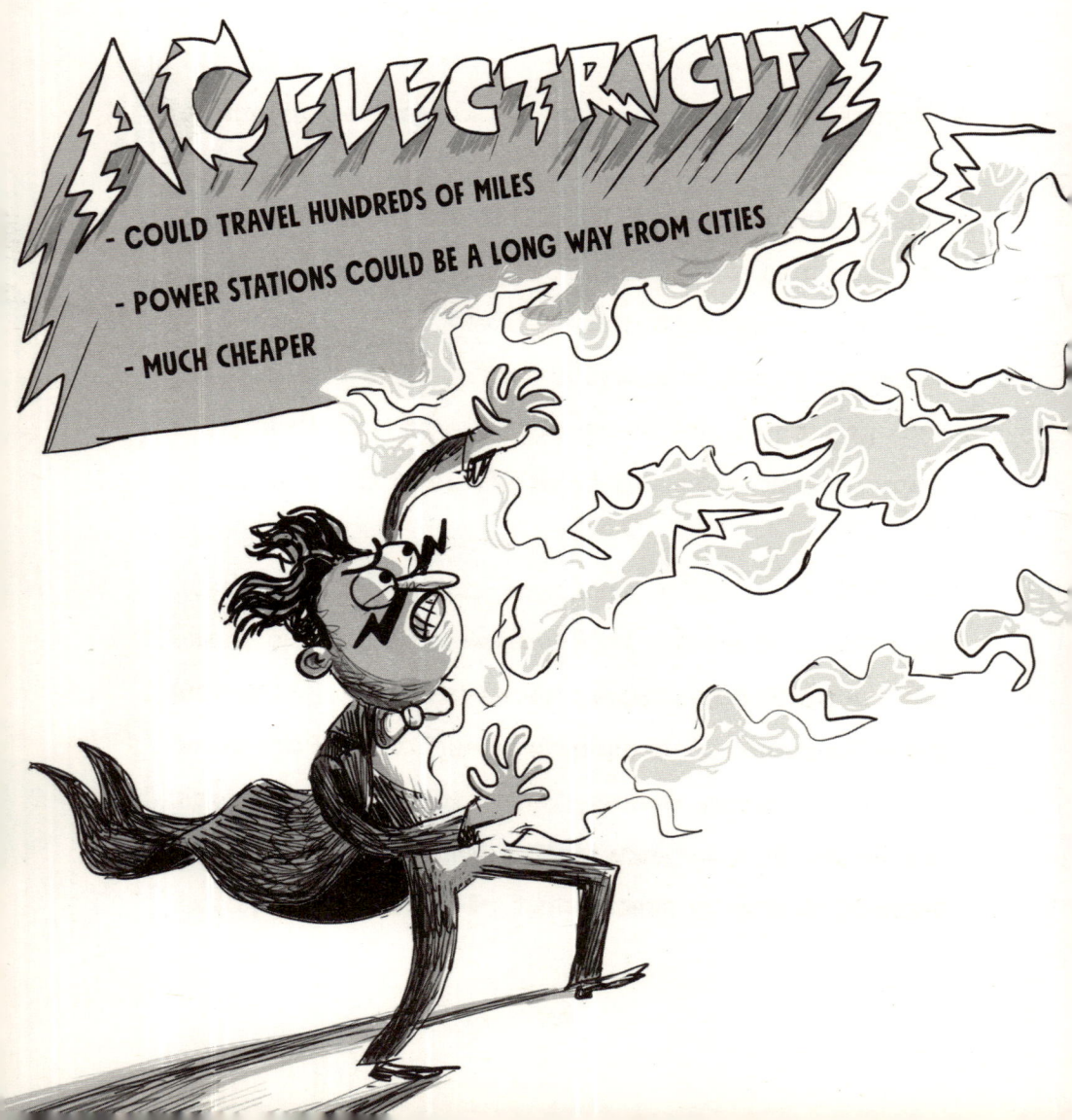

AC ELECTRICITY

- COULD TRAVEL HUNDREDS OF MILES
- POWER STATIONS COULD BE A LONG WAY FROM CITIES
- MUCH CHEAPER

DC ELECTRICITY

- COULD ONLY TRAVEL ABOUT ONE MILE
- THERE WOULD HAVE TO BE POLLUTING POWER STATIONS EVERY FEW STREETS
- MORE EXPENSIVE

As you can see, AC electricity was about fifteen million times better than DC. This annoyed Tommy quite a lot. But the thing that *really* annoyed him was that Nikola Tesla, his puny former employee, was a lot more popular than him. I know how he feels – my dog Pippin gets loads more fan mail than I do. ⚡ **Fact check – this is totally accurate.** ⚡

Tommy came up with a simple plan to get his dodgy old DC electricity into every house in the country – he decided to cheat. He started to spread gossip that AC electricity was extremely dangerous. And to prove that it was dangerous, he borrowed an elephant called Topsy from New York Zoo and . . . electrocuted it with AC electricity in front of a crowd of five hundred screaming people. I can't imagine the zoo was too happy about this either. Or Topsy, to be honest. (My lawyer, Nigel, has asked me to add that under no circumstances should you electrocute an elephant, or any other animal, even as part of a dastardly scheme.)

Edison's rotten plan didn't work though. The world decided that AC was best, and it's still the type of electricity that goes into our houses today. Even better, Tesla got a type of car named after him. That's right, the Lamborghinikola. **➤ Fact check – my data informs me that this is an unsuccessful attempt at a joke. ⚡**

It's time to start up my robot butler's lie detector and work out which of these facts about Nikola Tesla is a massive Fourth of July.

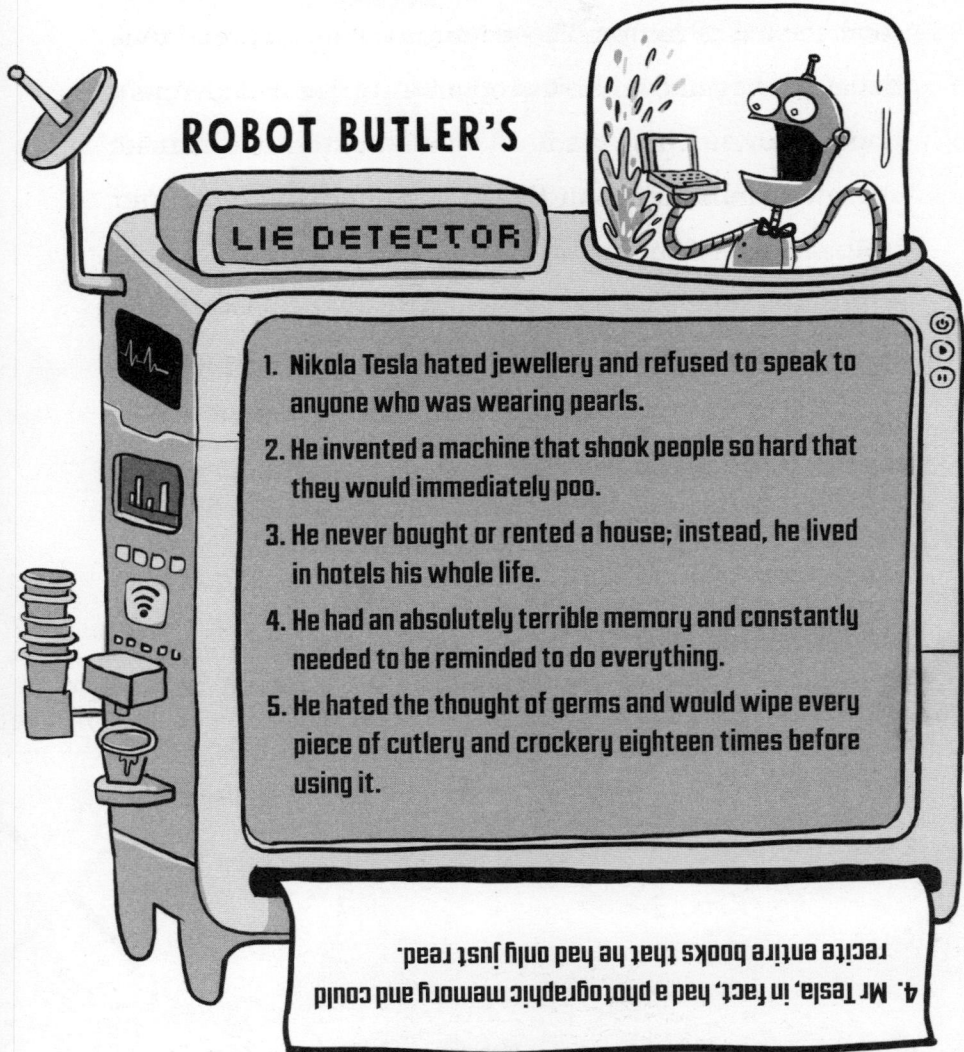

ROBOT BUTLER'S

LIE DETECTOR

1. Nikola Tesla hated jewellery and refused to speak to anyone who was wearing pearls.

2. He invented a machine that shook people so hard that they would immediately poo.

3. He never bought or rented a house; instead, he lived in hotels his whole life.

4. He had an absolutely terrible memory and constantly needed to be reminded to do everything.

5. He hated the thought of germs and would wipe every piece of cutlery and crockery eighteen times before using it.

4. Mr Tesla, in fact, had a photographic memory and could recite entire books that he had only just read.

LET THERE BE LIGHT BULBS!

Before electricity, people would light up their living rooms using oil lamps. These lamps weren't totally ideal, because they had a habit of constantly burning people's houses down. This was until a swan with an enormous scientist called Joseph Beard invented the first light bulb. Sorry – I mean, this was until a swan with an enormous beard called Joseph Scientist invented the first light bulb. Sorry again – I mean, this was until a scientist with an enormous beard called Joseph Swan invented the first light bulb.

He came up with the idea of a filament, which is a thin bit of wire that lights up when electricity passes through it. Joseph's filaments weren't perfect, because you had to get a new light bulb every time you turned it on, but his house was officially the first place in the world to be lit by electricity! That's quite a good claim to fame, but my house was the first place in the world to have Inflatable Cutlery from Adam Kay Genuis Enterprises Limited, which I think you'll agree is a lot more impressive.

IT AIN'T EASY BEING GREEN

Electricity changed our lives forever in all sorts of brilliant ways – from lights to laptops to radios to rollercoasters – it's basically impossible to imagine life without it. But it has also changed the world forever, and not in a good way. A lot of the electricity we use is produced by burning things like coal, gas and oil, which all have a terrible effect on the environment – they cause pollution, which heats up the planet.

But, hang on. Is it *really* that bad if the world's getting hotter? Doesn't it just mean we can throw away our scratchy old jumpers and eat more ice cream? I'm afraid it's actually not good news at all. If the weather gets hotter, it means that glaciers melt, which means sea levels rise and cause major flooding, plus animals like polar bears won't have anywhere to live and will start to die out. It also means there will be more droughts, more fires and more hurricanes. And eventually a lot of the crops we use for our food won't grow any more, and there will be whole countries that aren't safe to live in. Don't panic though – we can stop this from happening.

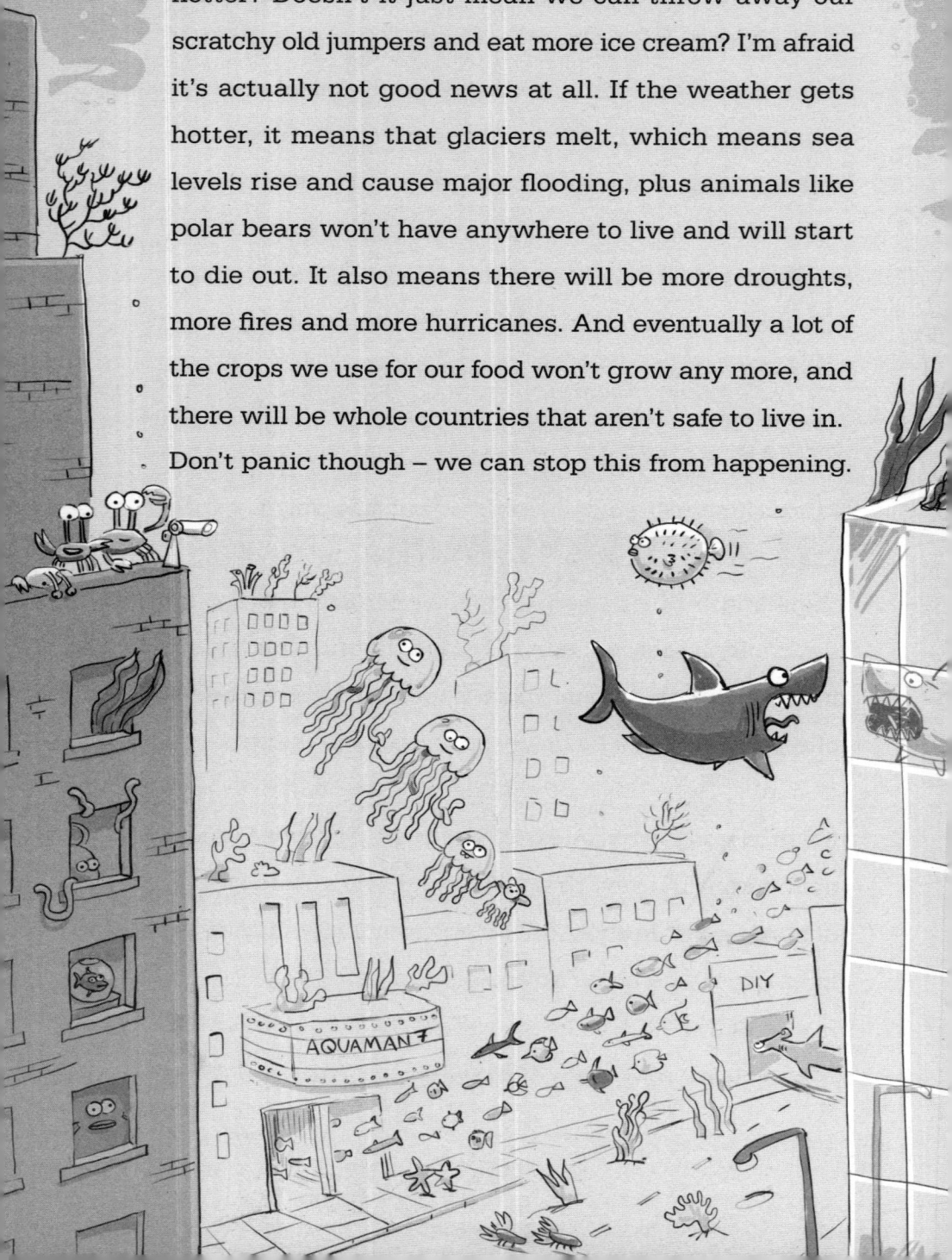

And we're not going to have to abandon electricity and go back to living in the dark, with no iPads, travelling around by horse. Electricity isn't the problem – the problem is the way we've been making it. There are lots of other ways to make electricity, and the environment will thank us a lot for using them. OK, it might not send you a thank-you card, but you won't have to live underwater, which is probably better than getting a thank-you card if you think about it.

SOLAR POWER

I don't know if you've noticed, but the sun is pretty bright. If we got solar panels to capture just one ten-thousandth of its energy, that would provide enough electricity for the entire world. One ten-thousandth isn't much – *these five words just here* are one ten-thousandth of this book.

Solar energy isn't a new idea – in 1948, a genius inventor called Mária Telkes designed and built a house totally heated by the sun. And it's not just buildings that can be powered using solar energy; it's aeroplanes too – a Swiss doctor called Bertrand Piccard invented a plane that was completely solar-powered. What's more, it

could fly for sixteen hours. In fact, in 2016, it was flown the entire way round the world! Bertie is from a famous family of explorers: his grandad Auguste and his Great Aunt Jeannette both broke records for flying into space in hot-air balloons; his dad, Jacques, was the first person to go to the bottom of the Mariana Trench, the deepest part of the world's oceans; and his uncle Jean-Luc Picard was captain of the starship *Enterprise*. ➤ **Fact check – Jean-Luc Picard is a fictional character, but was named after Auguste Piccard.** ➤

WIND

We could also power the entire planet just using wind. I don't mean farting, so no energy companies will be employing you or Pippin any time soon, I'm afraid – I'm talking about wind turbines: those tall, spindly

windmills that you might have seen in the countryside and out at sea. People have been using wind power to do things like smoosh up wheat for more than a thousand years, but the first-ever turbine that made electricity was built in 1887 by an American inventor with the amazing name of Charles Brush. Even though wind energy is free, it's quite expensive building these turbines – if you want a nice big turbine in your garden, that'll be about two million pounds, please.

WATER

Princes don't do very much these days, do they? They're mostly wearing crowns, opening new supermarkets and writing the occasional terrible book. But one hundred years ago a prince in Italy called Prince Piero Ginori Conti was the first person to generate electricity using water power. His machine was bigger than a car and only managed to light four light bulbs, but we've all got to begin somewhere. For example, this book started out as a single sentence and now it's the best book in the world. ➤ **Fact check – better books than this one include, in alphabetical order: *Aardvark Anatomy, Aaron's Abacus* –** ➤ Right, let's move on.

TRUE OR POO?

ONE OF THE FIRST-EVER BATTERIES WAS MADE OUT OF FROGS' LEGS.

TRUE If you're a frog, I'd stop reading right now. In 1845, a man called Carlo Matteucci discovered that if you cut the legs off ten frogs, then join them together, they create a small electric current. Battery technology has obviously moved on a lot since then, which is a relief – I'm glad I don't have to cut up a load of frogs every time the remote control stops working. Frogs, you can start reading again now.

EARLY LIGHT-BULB FILAMENTS WERE MADE OF NOSE HAIR.

POO Thomas Edison experimented with all sorts of materials to use as filaments in his light bulbs. Bits of cork, coconut hair, silk fibres and even hair plucked from some of his employees' beards – I think I'd have complained to Human Resources if my boss did that to me. But he never used nose hair, I'm pleased to say.

ENGINEERS ARE CURRENTLY BUILDING WIND TURBINES IN SPACE.

POO Unfortunately, there isn't any wind in space, except for the occasional alien fart from one of the Octopus People of Zaarg. But there are plans to float some turbines on wires a couple of miles above the ground, where the wind is a lot faster. Like massive spinny kites.

ADAM'S ANSWERS

HOW MUCH WAS THE WORLD'S MOST EXPENSIVE BATTERY?

Next time a grown-up complains about how much money they're spending on batteries for all your gadgets, tell them they're lucky that your batteries are so cheap. The Crimson Storage battery in California, which was built to provide energy to fifty thousand houses, cost about half a billion pounds and was the size of a thousand football pitches, so it probably wouldn't fit in the drawer where you keep the batteries.

I DON'T THINK THAT'S GOING TO FIT IN THE BATTERY DRAWER.

WHAT'S THE LONGEST A LIGHT BULB HAS LASTED FOR?

Oh, just a million hours or so. A light bulb in a fire station in America has been on pretty much constantly for over a hundred and twenty years. It used to be extremely bright, but it's so old that it's pretty dim these days. A bit like my Great Aunt Prunella.

HOW DOES BURNING COAL PRODUCE ELECTRICITY?

Good question, me. Thanks, me. Well, if you burn coal, then it gets very hot. Then the heat boils water. Then the water turns to steam. Then the steam turns a wheel. Then the wheel spins a magnet through a tube of copper . . . which is exactly how Fartaday worked out how to make electricity all those years ago.

EXPENSIONS

Sometimes being an inventor is a costly business. I spent nearly a thousand pounds developing Soup Paint. (Available now from Adam Kay Genuis Enterprises Limited – paint your bedroom with Soup Paint and if you get hungry during the night all you need to do is lick the walls!) But some inventions have cost even more than that. Here are some of the priciest products ever produced.

THE CHANNEL TUNNEL

In 1994, the Channel Tunnel opened, and people could take a train from England to France for the first time. This was possible because of some massive tunnel-boring machines, which are so dull they make the ground yawn open so you can build a tunnel. ➤ **Fact check – 'boring' means 'digging'.** ➤ They started digging from both ends at the same time, and must have been extremely relieved when they joined up correctly in the middle. One of the boring machines has been left buried under the Channel because nobody could be bothered driving it out again. Five hundred million people have travelled through the tunnel, plus over two million cats and dogs. Including Pippin, who once did a massive poo on the train, then started eating it. The poo, that is – not the train. It took thirteen thousand people six years to build the Channel Tunnel, and it cost over five billion pounds. You could buy a lot of croissants for that.

THE INTERNATIONAL SPACE STATION

It's the size of a football pitch. It flies 250 miles above Earth. That's right – it's the Amazing Space Football

Pitch! 🠖 **Fact check - it's the International Space Station.** 🠔 Oh yeah. The International Space Station, or ISS to its friends, is a floating laboratory that astronauts use to investigate all things space-related, such as: 'Are there any aliens?' 🠖 **Fact check - none that we have currently identified.** 🠔 and 'Can you eat soup in zero gravity?' 🠖 **Fact check - yes.** 🠔 Seven astronauts live there at all times and whizz round Earth at 17,000 miles per hour, which is over thirty times faster than an aeroplane. If you want to build your own International Space Station, you'd better save up your pocket money because it cost over 125 billion pounds. Oh, and it costs about a million pounds a day to keep it going, so make sure you budget for that as well. The ISS has had astronauts on board carrying out experiments continuously since the year 2000 . . . Well, not the same astronauts the whole time: you can't spend that long in space, otherwise when would you go to the dentist for a check-up?

TOILET

How much do you think the world's most expensive toilet cost? Nope, more than that. Still more. Now multiply it by a hundred. The world's most expensive toilet cost five million pounds. It was built by Maurizio

Cattelan out of solid gold and didn't even have a heated seat – so it must have been extremely cold and unpleasant for the old bum. Want to use it? Well, I'm afraid you can't – a thief stole it in 2019. I guess that's the risk if you have a five-million-pound toilet.

HEY! I'M STILL USING IT!

THE GREAT MOSQUE OF MECCA

This isn't just the most sacred place in the religion of Islam and the biggest mosque in the world – it's also the most expensive building ever constructed. It's the size of one hundred football pitches! It can hold two million people – that's more people than live in Birmingham and Manchester combined! And it's got thirteen thousand toilets! And how much did it cost? Just the small matter of one hundred billion dollars.

PART TWO

OUT AND ABOUT

SUBMARINES, SATNAV AND SPACE TOILETS

SATNAV'S BROKEN AGAIN.

IS THERE LIFE ON EARTH?

BUILDING

If you're a fan of things like living inside, going on roads, crossing rivers using bridges, driving through tunnels and being able to flush the toilet, then this is the chapter for you. If you live outside, poo in the woods and never use roads, bridges or tunnels, then you might not find it so interesting. **➤ Fact check – according to my data there are over seven hundred other reasons why people would not find this chapter interesting. ➤**

HENGEINEERING

I do love visiting a nice henge. What's your favourite henge? For me it's a tough choice between Stonehenge and Bumhenge. **➤ Fact check – Bumhenge is a formation of pebbles in the shape of a buttock that you placed on your windowsill.➤** Stonehenge is a circle of absolutely massive stones on Salisbury Plain, in the south of England. It was built about five thousand years ago and no one is totally sure why. Some people think it was built for religious reasons, some people think it's to do with the sun, and others think it was a kind of ancient hospital. I think it was built just to confuse people like us thousands of years later.

But one thing we know for sure is that building it would have been an absolutely enormous faff. The stones weigh up to 25 tonnes, which is the same as a double-decker bus with four hippos in it, and they had to be brought from over twenty miles away, and then pulled upright. It's incredible to think that this was done just by people dragging them on the ground and lifting them with ropes, because I don't think there were such things as cranes and lorries back then. ➤ **Fact check – you are correct for the first time in 26 pages.** ➤

OH DEAR. HE'S THE ONLY PERSON WHO KNEW WHY WE'RE BUILDING THIS.

A few years later, over in Egypt, the people there built something far more impressive. (No offence, Hengey.) The Great Pyramid of Giza is made out of more than two million huge blocks of stone, it's taller than Big Ben, and it weighs more than five thousand jumbo jets. The pyramids in Egypt were built as tombs for their dead pharaohs – the Great Pyramid of Giza was made for a giza called Khufu. Something tells me that Khufu might have been a tiny bit of a show-off. When it was built, the Great Pyramid was the tallest building in the world, and no one built anything taller for nearly four thousand years. Also, it hasn't fallen down. Unlike Stonehenge. (No offence again.)

HOW'S THE MASSIVE STATUE OF ME GOING?

WELL, WE DID YOUR NOSE THEN RAN OUT OF BRICKS.

The Great Pyramid was one of the Seven Wonders of the Ancient World: amazing feats of engineering from thousands of years ago. In fact, it's the only one that's still standing. See if you can guess which the other six were, and sort the wonders from the blunders:

A) The Banging Gardens of Basildon

B) The Colossus of Rhodes

C) The Turd of Toulouse

D) The Leaning Tower of Pisa

E) The Statue of Zeus at Olympia

F) The Mausoleum at Halicarnassus

G) The Bread Section of Sainsbury's

H) The Lighthouse of Alexandria

I) The Gift Shop of Blackpool

J) The Hanging Gardens of Babylon

K) The Temple of Artemis

If you chose B, E, F, H, J and K, then you're a wonder of the world. If you chose A, C, D, G or I, then you're a bumspot of the bin. Although I personally feel that the Bread Section of Sainsbury's was cruelly overlooked for a place on the list.

GET OVER IT

The first bridges were pretty basic. If a tree fell down across a river, someone would walk across it and say, 'Cool bridge!' Later, people would drag trees over deliberately or, if there weren't any trees around, they'd use things like stepping stones. But these methods weren't great if you were carrying loads of shopping or moving house, so people had to get a bit more creative and started building bridges out of stone. These early bridges were made by building an arch from one side of the river to the other, and then making a straight path on top of it.

The bridge over the River Meles in Turkey is in the *Guinness Book of Records* for the most people dancing on it dressed as a polar bear. **Fact check – it has the record for being the oldest bridge still in use, dating from nearly three thousand years ago.** The Romans really nailed the whole bridge-building thing, and loads of their bridges are still standing today. They were built so well because Roman bridge-builders didn't get paid for their work until years and years after the bridge was finished, just in case they fell down.

These days, lots of bridges are what's known as suspension bridges. You might have seen them before: they look like they're held up by bits of string tied to some poles. You'll be relieved to hear that no string is involved, and the roads are actually held up by strong

metal cables that are attached to tall towers and then buried deep into the ground. This makes the bridges very secure and means that they can stretch across much longer distances than the old-school arch bridges. The first metal suspension bridge was built in Pennsylvania in 1801 by Jeff Bridges. ➤ **Fact check – the bridge was constructed by James Finley. Jeff Bridges is an old movie star who 0.04% of your readers have heard of.** ➤ One of the most famous suspension bridges in the world is the Brooklyn Bridge in New York. It was designed by John Roebling in 1867, but then he died (oops), so his son Washington took over, but then he got ill (oops again), so Washington's wife, Emily Roebling, took over. And she absolutely aced it. She was the first person to cross the bridge, which she did carrying a rooster. And why not.

THIS IS EVEN BETTER THAN THE DAY I CROSSED THE ROAD!

Britain's biggest bridge boffin was a man called Isambard Kingdom Brunel. Ever crossed the Clifton Suspension Bridge in Bristol? That's Isambard! Ever taken the Royal Albert Bridge between Devon and Cornwall? That's Isambard! Ever sailed on the huge steamship the *Great Western*? That's unlikely, because it was smashed up in 1856.

Isambard (why does no one call their children Isambard any more?) was born in 1806 and became one of the country's leading engineers – building bridges, ships and railways all over the place. He was great mates with Sarah Guppy, who was the person who invented the automatic breakfast machine, and she worked with Izzy on a bunch of his projects. IKB was voted number two of the Top 100 Greatest Britons of all time, just behind me. ➤ Fact check – just behind Winston Churchill. ➤

It's time to turn on my robot butler's lie detector to see if you can work out which of these facts about Isambard Kingdom Brunel is a total samurai.

ROBOT BUTLER'S

LIE DETECTOR

1. Isambard Kingdom Brunel's mother was sent to prison for being a spy.
2. He was extremely tall and needed special doorways in his house so he didn't knock his head.
3. He nearly died when he was performing a magic trick for his children and got a coin stuck in his windpipe.
4. His first job was as a clockmaker.
5. Workers on Isambard's projects were more likely to die than soldiers who fought in the First World War.

2. Mr Brunel was, in fact, rather short and wore huge hats to try to make himself look taller.

(PS: My lawyer, Nigel, has asked me to point out that you should never perform any magic tricks that involve getting a coin stuck in your windpipe.)

GOING UNDERGROUND

If, like me, you're a bit scared of heights, you might not be a massive fan of bridges. Particularly not Haohan Qiao in China, which translates as 'Brave Man's Bridge' because it goes between two cliffs and the walkway is totally made out of glass. No. Thank. You. But why go over when you can go under? Please welcome to the stage . . . tunnels!

Tunnels have been around for even longer than my Great Aunt Prunella. In fact, they've existed since the first caveperson thought their cave wasn't quite big enough and grabbed a pickaxe. I guess conservatories hadn't been invented yet. About three thousand years ago, in Ancient Persia (now Iran), they built qanats, which were tunnels to transport fresh water from lakes and wells to

the people who needed it. And because tunnels are underground ➤ **Fact check – this is a basic fact that all humans over the age of three know.** ➤ it meant that the water wouldn't evaporate. The qanats had to be chipped away by hand, and the workers would often only manage to get through five millimetres a day, which is as long as this. ➤ **Fact check – unclear. As long as what?** ➤ This. The word 'this'.

Eventually, tunnel-diggers realized that if you wanted to make a hole in some rock, you could just blow it up. About five hundred years ago, they started using gunpowder to smash holes in rock, which didn't cause any problems at all, unless you count people getting accidentally exploded by the gunpowder, in which case it caused quite a lot of problems.

In 1807, a champion wrestler from Cornwall called Richard Trevithick started to build a tunnel under the River Thames. Unfortunately, the tunnel collapsed and flooded, almost killing him. He probably should have stuck to the wrestling. Sixteen years later, an engineer called Marc Brunel thought he'd have a go. That's a coincidence – he's got the same surname as Isambard! ➤ **Fact check - he was Isambard Kingdom Brunel's dad.** ➤ Well, Daddy Brunes had a clever idea to stop the tunnel collapsing, which he called a tunnelling shield. It was basically a big machine that pushed forward and held the roof up. It must have looked pretty exciting because hundreds of people came every day and paid money to watch it work. But I guess there was no such thing as telly back then. The Thames tunnel (between Rotherhithe and Wapping, tunnel fans) finally opened in 1843, and it became the most popular tourist attraction in the world, with over a million people walking through it in the first three months. Like I say, no telly. You can still go through PapaBru's tunnel today if you take the East London line in London. ➤ **Fact check - less than half of the London Underground is actually underground.** ➤

About twenty years later, a man called Alfred Nobel found an even more effective way of blasting tunnels. In 1867, he mixed a dangerous chemical called nitroglycerine with some silica from those little packets that come with your shoes that say 'silica gel – do not eat' ⚡ **Fact check - silica comes from sand.** ⚡ then he shoved it into some sticks, and wrote 'DYNAMITE' on the side. And do you know what he called it . . . Oh, how did you guess?

URIN SCORE
7/10
GOOD NAME,
WHICH COMES
FROM THE
GREEK WORD
'DYNAMIS',
MEANING
'POWER'.

Alf got really upset that people weren't building tunnels with his lovely new invention – most people were using it in wars to blow people up. He used his dyna-money to start some special prizes to celebrate brilliant scientists and writers who improve the world rather than explode it. They're basically like the Oscars for scientists, except you don't get to wear such good outfits. So far I have won fifteen Nobel Prizes. ➤ **Fact check – the only prize you have won is printed below.** ⚡

ROADS OF FUN

Where were roads invented?

A) Um

B) Ur

C) I don't know.

The correct answer was B – if you got that right, then you are allowed to eat twenty-seven tubs of ice cream. (My lawyer, Nigel, has asked me to point out that you are absolutely not allowed to do this.) The first road was built in a place called Ur, in what's now Iraq, about six thousand years ago. It was built out of mud bricks, which are bricks made out of badger poo. ➤ **Fact check – they're made out of mud.** ➤ Some of the bricks even have doggy footprints in, from a naughty dog who walked on them before they were dry. Probably Pippin's great-great-great-great-great-great-great-great-great-great-great-great-great-great-granddog.

The Romans really zhuzhed up the road system when they came along. They built their roads from lots and lots of different-sized stones, which would squash down to make a nice smooth surface, and – like their bridges – a lot of their roads are still around today. They also worked out that if a road slopes at its sides, then all the water will slide off and it won't be so likely to flood. They even came up with the idea of putting signs along the road to say how far they were from the nearest city, which stopped all the Roman *pueri* and *puellae* in the back of the chariots asking, 'Are we nearly there yet?' **Fact check – 'pueri' and 'puellae' mean 'boys' and 'girls'. It is confusing that you understand Latin better than English.**

MOTORWAYS

The first motorways in the world were opened in Italy and Germany back in 1924, when they got bored with driving slowly everywhere. Britain's first three major motorways opened on the same day in November 1959: the M1 and – can you guess what the other two were called? That's right, the M10 and the M45.

URIN SCORE
3/10
INABILITY
TO COUNT.

CATSEYES

If you've ever been on a road at night, you'll know that the different lanes are separated by little light-up studs. They're called Catseyes, and they were invented by Percy Shaw in 1933. Perce was driving along one foggy night and couldn't see where he was going, until he saw his headlights reflected in a cat's eyes and thought that would be an excellent way of helping drivers. So he designed these little rubber things with glass tubes inside that go in the road and reflect any light that hits them, keeping drivers safe.

WHY DIDN'T YOU MENTION MY NAME?

PETER FUDGINGTON HAIRBALL THE 3RD

There are over half a billion Catseyes in the UK alone. It's a shame he didn't see his headlights reflected in a cow's bum, because then there would be half a billion Cowsbums in the UK, which would be much funnier.

ZEBRA CROSSINGS

URIN SCORE
8/10
MAKES STREETS
SEEM EXCITING.

The first-ever zebra crossing was painted on a road in Slough, near London, in 1951, to make it safer for pedestrians to cross the road. It was given its name by a bloke called James Callaghan, who would later become the Prime Minister of the UK. He went to see it and thought that its stripes looked a bit zebra-ish. Other types of crossing include the pelican crossing (with big light-up lollipop poles at either end), the toucan crossing (for both people and cyclists because 'two can' cross), the Pegasus crossing (for horses) **Fact check – you are not lying.** and the Bigfoot crossing (for enormous monkeys). **Fact check – you are now lying.**

TALL STOREYS

If you're in the middle of a city, and you need a huge new office but there's absolutely no space left on the ground, what do you do? That's right – you build an enormous underground lair. **➤ Fact check – you build up into the sky.➤** Really tall buildings have only been possible in the last hundred and fifty years because of one big issue: laziness. It wasn't the builders being lazy about making them; it was the people who used them. No one could be bothered climbing 342 flights of stairs – can you blame them? – so buildings were only a few storeys high until a clever new invention came along. Jetpacks. **➤ Fact check – lifts.➤** There had been a few lifts made in the past: for example, a 'flying chair'

that King Louis XV had installed in one of his palaces in 1743 so he could get up to his balcony without bothering with pesky stairs. But it had to be cranked by hand, which must have been extremely annoying.

THIS . . . IS . . . SO . . . CONVENIENT . . .

People started to come up with different ways of lifting the lifts (or elevating the elevators, as they say in America), such as using steam power or high-pressure water, but the real game changer came in 1880 when Werner von Siemens came up with the first electric lift in Germany.

Even though being in a lift means you're dangling in a box held up by a bit of rope, lifts are extremely safe, thanks to a man called Elisha Otis, who designed a safety mechanism in the 1850s that stops it falling if the rope breaks. He demonstrated his new lift by standing in it while a man with an axe cut through the cable . . . and the lift stayed where it was. Otis's company, called Otis, still make lifts now – in fact, more than two billion people travel in them every day. I don't think they do the thing with the axeman any more, you'll be pleased to hear.

The first-ever skyscraper was the Home Insurance Building, built in Chicago in 1885, but it was only ten storeys high so it didn't really scrape anything. These days skyscrapers are some of the most famous buildings in the world, including the Empire State Building and the Chrysler Building in New York, and the Shard and the Gherkin in London. The battle to build the biggest skyscraper has been going on for ages, and people are always building new ones. At the time of writing this, the tallest building in the world is the Burj Khalifa in Dubai – it's over 160 storeys high, it's got the highest restaurant in the world, and to climb to the top using the stairs would take people more than half an hour. ⌁ **Fact check – the data I have collected on your health suggests you would collapse in a heap after two minutes and nine seconds.** ⌁ In the next thirty years, it's likely that the tallest skyscrapers will be over a mile high. Until they're knocked down by the Octopus People of Zaarg, that is.

THE GHERKIN

THE CHERRY TOMATO

THE MANKY OLD PIECE OF CHEESE

THE SHARD

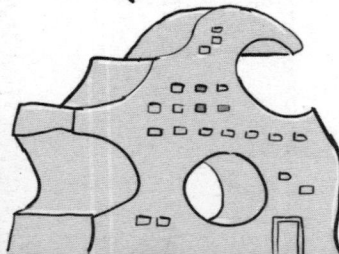
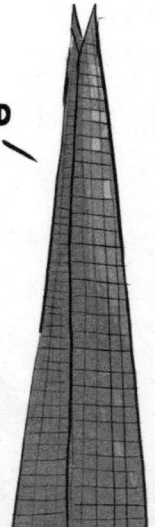

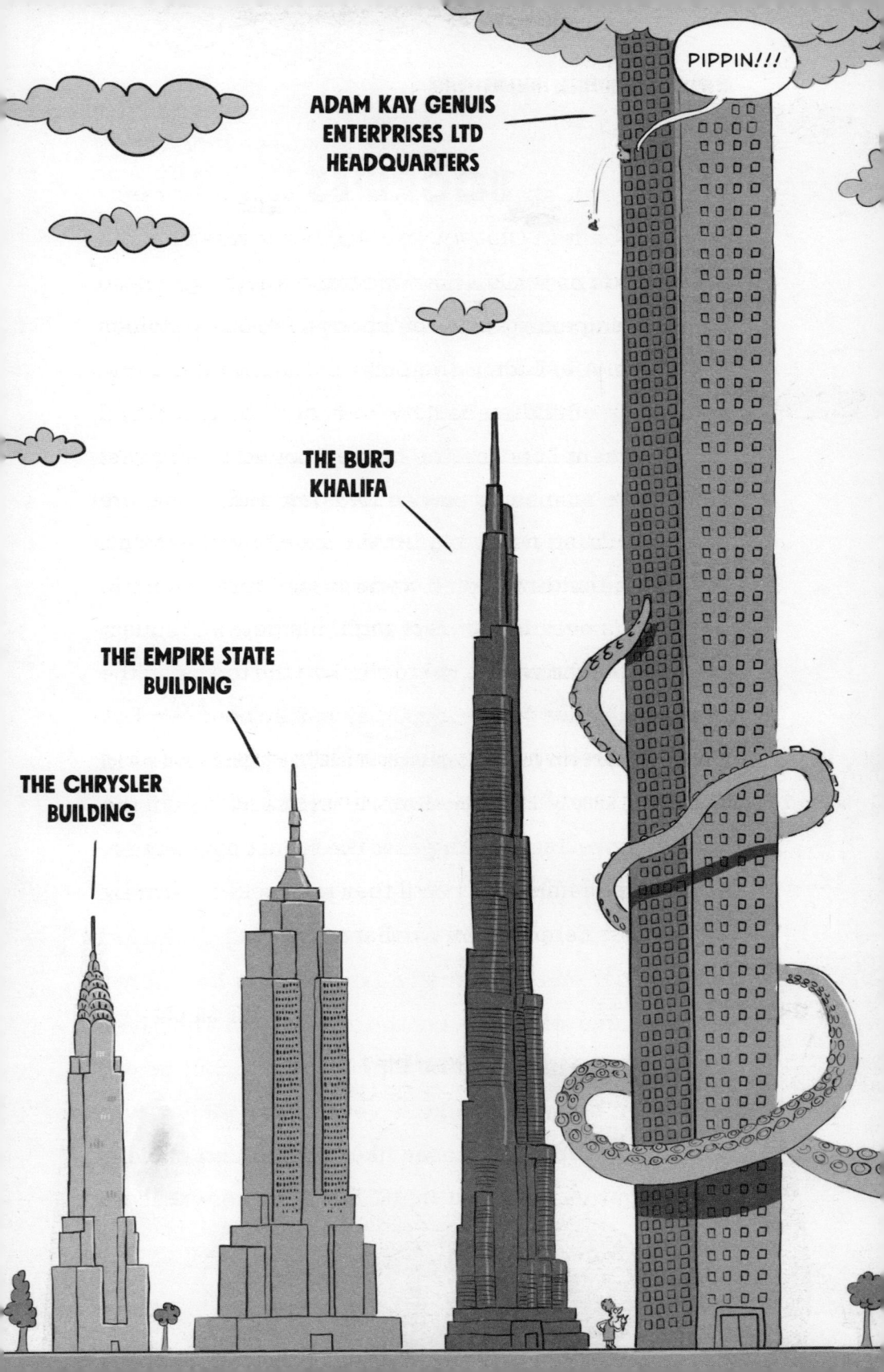

BOGPIPES

Your body is basically a machine that turns food into poo. I'm not being rude – everyone's body is. Yours, Beyoncé's, the King's. And because you make 150 kilograms of poo every year, which is about the weight of a large panda, it's important that there's somewhere for it all to go. And that's thanks to sewers. No, not people who do sewing. I mean those big underground pipes that our toilets drain into. You'll remember them from the bathroom section. ⚡ **Fact check – there is a 3% chance anyone will remember them.** ⚡

The first system of sewers was built five thousand years ago in the city of Mohenjo-Daro, in what's now Pakistan. London took a little bit longer to catch on, unfortunately. Londoners preferred to poo into big holes in the ground called cesspits, and then, when those got full, they'd just dig another one. Eventually, London was basically a massive sea of poo – on the streets, in the river, everywhere. This might be Pippin's dream, but no one back then seemed to like it very much, plus it was spreading horrible diseases like typhoid and cholera. Things got even worse in 1858 when there was an extremely hot summer, known as the Great Stink, and London started to pong so badly that people were fainting in the streets. Probably onto a big pile of poo.

Parliament decided that London needed a proper sewer system and instructed it to be built by Joseph Bazalgette, who was famous for inventing the baguette. ➤ **Fact check – he was a famous engineer.** ➤ JoBaz built more than a thousand miles of sewers underneath London, which we're still using today.

HEY! NO LITTERING!

Don't read this next bit unless you've fully digested your last meal. Are you ready? Here we go.

Today, our sewers are now clogged up with fatbergs. Fatbergs are absolutely enormous rock-solid balls of cooking fat, wet wipes and nappies, caused by people pouring oil down their sinks and flushing away things that should go in the bin. The biggest-ever fatberg was found in London in 2017 – this massive heap of yuck was the length of two football pitches and took a team of eight people three solid weeks to blast away.

TRUE OR POO?

IN SPAIN, THERE ARE COW CROSSINGS WITH SPOTS INSTEAD OF ZEBRA CROSSINGS.

TRUE The city of A Coruña produces more milk than anywhere else in Spain, thanks to its one million cows. The citizens were so proud of their cows that they painted blobs on the ground so you now cross the road using cow crossings. I think that was a very sensible mooooove.

ERMINTRUDE!

THE EIFFEL TOWER CHANGES IN SIZE.

TRUE You know the Eiffel Tower? It looks like a huge electricity pylon plonked in the middle of Paris, and was built in 1887 by an engineer called Gustave Eiffel (whose surname rhymed with 'trifle' and who helped design the Statue of Liberty). Anyway, all metal expands when it gets hotter – not so much that you'd notice your keys getting bigger, but just by a tiny fraction of a fraction of 1%. Because the Eiffel Tower is so massive, even that tiny fraction means that in summer it grows by about fifteen centimetres compared with winter. That's about the length of a tube of toothpaste.

THE FORTH BRIDGE BETWEEN EDINBURGH AND FIFE TAKES SO LONG TO PAINT THAT AS SOON AS IT'S FINISHED, THEY HAVE TO GO BACK AND START ALL OVER AGAIN.

POO Lots of people think this is true, but it's actually never been the case. In fact, the last coat of paint it was given was a special kind that doesn't need to be redone for another thirty years.

ADAM'S ANSWERS

WHAT'S THE LONGEST TUNNEL IN THE WORLD YOU CAN DRIVE THROUGH?

The Lærdal Tunnel in Norway is more than twenty-four kilometres long, so it would take you about five hours to walk through it. If you piled up the amount of rock they dug out to make the tunnel, it would be nearly three times as big as the Empire State Building.

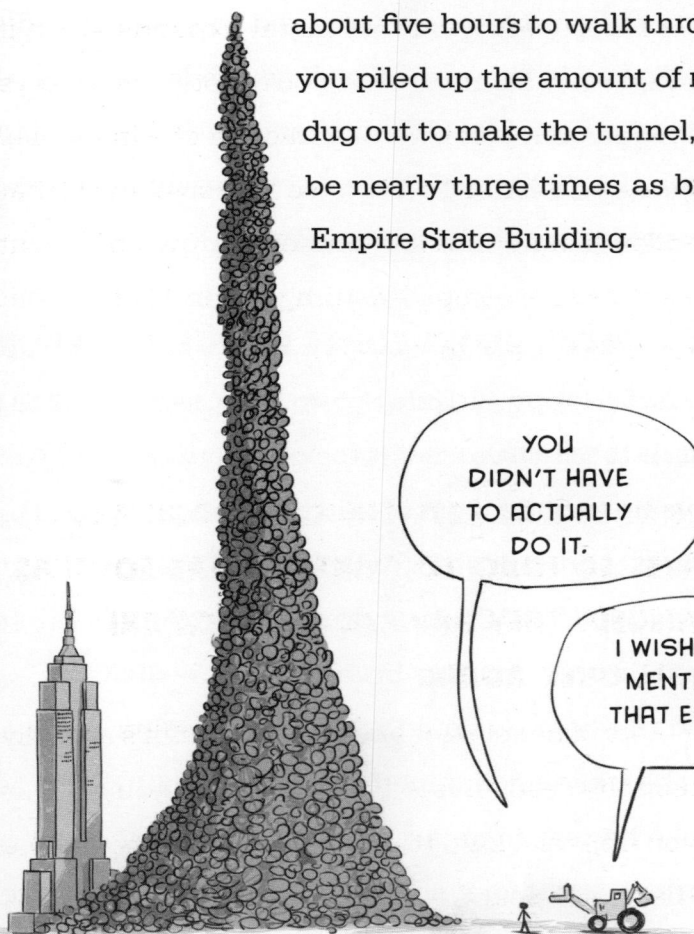

YOU DIDN'T HAVE TO ACTUALLY DO IT.

I WISH YOU'D MENTIONED THAT EARLIER.

WHEN DID WE START USING TRAFFIC LIGHTS?

The first-ever traffic lights were installed next to the Houses of Parliament in 1868 and were made out of two red and green gas lamps. They weren't very popular for two reasons. Firstly, they weren't automatic, so a police officer had to stand there twenty-four hours a day and turn them from red to green to red to green to red to green. Secondly, they occasionally exploded and injured the police officer. The UK's first automatic traffic lights appeared in 1927 in Wolverhampton, using the much less explodey method of electricity.

WHAT'S SPECIAL ABOUT TOWER BRIDGE IN LONDON?

It can fart 'Happy Birthday'. **Fact check – it lifts up to allow boats to sail through.** It took four hundred workers eight years to build it out of thirty million bricks and ten thousand tonnes of steel, which is the weight of sixty blue whales. In 1952, a bus driver called Albert Gunter was driving across the bridge when – disaster – it started to open! He couldn't turn back so he drove as fast as he could and jumped over the gap onto the other side, like he was in an action movie. In fact, I might go and write a script for it now. See you in the next chapter.

SPORTVENTIONS

Humans have been playing sport for a very long time – we know that cave dwellers were racing each other and having wrestling matches over fifteen thousand years ago, by looking at their Instagram accounts. #cavewrestling #winner #blessed ⚡ **Fact check – we know this by looking at cave paintings.** ⚡

THE OLYMPICS

The biggest sporting event in the olden days was the Olympics. They started about three thousand years ago and would happen every four years at a place called Olympia in Greece. Olympia – that's a coincidence! ➤ Fact check – it is not a coincidence. The Olympics are named after Olympia. ➤ The games were quite similar to the ones we have today, with a load of different sports being played, such as boxing, horse riding, running and throwing the javelin. One big difference was that everyone competed naked, because they thought you could run faster with no clothes on. This must have

made the javelin matches slightly nerve-wracking. The Ancient Olympics continued for a few hundred years, but then they got bored of them or forgot about them or someone complained about the 'no clothes during javelin' rule. The Olympics were brought back in 1896 by a man called Pierre de Coubertin, and they're still going today.

The Winter Olympics were added in 1924 to give penguins a chance at winning a medal **Fact check – sigh** and the Paralympics for athletes with disabilities began in 1960. The Summer Olympics feature thirty-

three different sports, although hopefully it will soon be thirty-four when they accept my suggestion of Competitive Poo Hurling.

FOOTBALL

Four thousand years ago in China there was a game called tsu chu, where you had to get a ball into the net without using your hands. Sound familiar? In Central America they had a version of the game where they wrapped the ball in cloth, soaked it in oil, then set it on fire, so you were playing with a literal fireball. I think I'd fake a letter from my mum excusing me from PE if I had to play that at school. In 1838, a bunch of people sat down and decided on some rules for the game that everyone should follow. The Cambridge Rules are the basis of how football is still played and include things like the teams wearing different colours, when throw-ins and goal kicks are taken, and that West Ham United are the best. ➤ **Fact check - it appears that you have added in a new rule here.** ➤

BASKETBALL

I always used to complain at school when we had to play sports outside in the winter, but the teachers would still make me do it, until my fingers all got frostbite and my nose fell off. ➤ **Fact check - this is not true.** ➤ Well, I clearly wasn't as good at moaning as a class of students in America back in 1891, who were being taught PE by James Naismith. He invented a game that would keep his students fit and healthy but that they could play inside – basketball! Instead of nets like we have now, he nailed some baskets that used to store peaches high up on the walls. And when everyone got bored with climbing up ladders to fetch the ball every time one went in, he cut the bottoms off the baskets.

TRAINERS

Until the 1870s, if you wanted to play sports you'd do it either in bare feet or in the uncomfortable leather shoes you wore to school or work. Finally, people realized that shoes made of cloth with a nice rubber sole might mean you could run a bit faster without your feet turning into mincemeat. In 1917, a man called Marquis Converse invented shoes specially for basketball players, called . . . you guessed it – Nike. **Fact check – they were called Converse.** Talking of Nike, they owe their success to a waffle-making machine. Bill Bowerman, one of the founders of Nike, was looking for a way to make his trainers grip better, and eventually poured some liquid rubber onto his wife's waffle-maker. The rubber set to give a criss-cross pattern, which he used on the soles of his trainers. The company's slogan 'Just do it' came from Mrs Bowerman's reply to Bill when he refused to go to the shop and buy her a new waffle-maker. **Fact check – no, it did not.**

ALL I NEED IS A LOGO. I'M SURE IT'S RIGHT UNDER MY NOSE . . .

NEW FROM

ADAM KAY GENUIS ENTERPRISES LIMITED

ADAM'S TREMENDOUS TEMPERATURE-CONTROLLED T-SHIRT

You know what it's like – you leave the house on a bright sunny day wearing a T-shirt and ten minutes later it's snowing. What you need is Adam's Tremendous Temperature-Controlled T-Shirt, which will automatically extend its sleeves to cover your arms and thicken up until it's a fleece.*

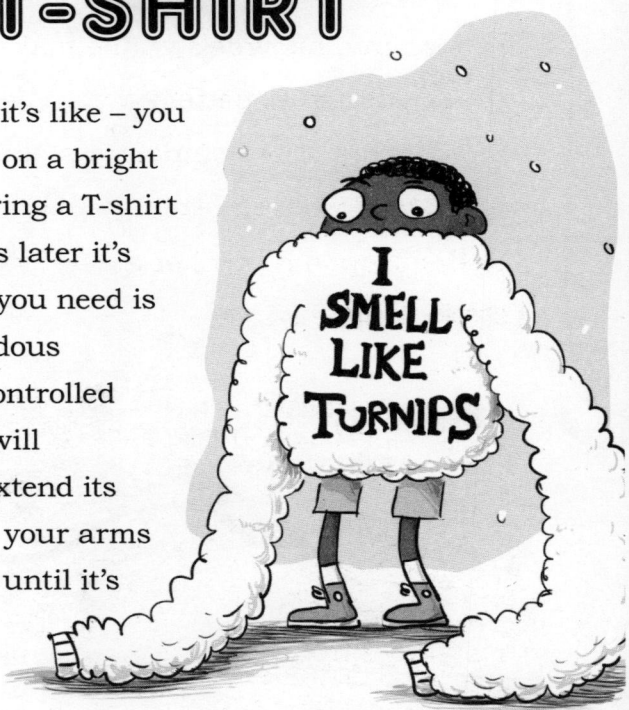

I SMELL LIKE TURNIPS

Only £3,243.99 (requires 48 AA batteries)
*Please note that the T-shirt only comes in one design, with 'I SMELL LIKE TURNIPS' printed in extremely large letters on the front and back.

TRAVEL:
ON THE GROUND

It's weird to think that a long time ago we all had to walk everywhere. Humans were around for hundreds of thousands of years before anyone built a bus or a boat or a bumcopter. **Fact check – the bumcopter does not exist.** Walking was slow, and pretty dangerous, depending on how many tigers and sabre-toothed porcupines were hanging around. **Fact check – sabre-toothed porcupines never existed either.** Over the next few chapters, we'll find out who we have to thank for the ways we now travel around.

ON THE RIGHT TRACK

The first-ever train tracks were built over two thousand years ago in Ancient Greece. In fact, the last time I was on a train I bought a sandwich from the buffet car that tasted like it had been there since back then. You couldn't actually travel on those trains; they were little trucks on rails pulled by horses, designed for moving stuff around.

If you wanted to get on a train yourself, you would have to wait at the platform for nearly two thousand years (and hope there weren't any delays on the line). The train everyone was waiting for was designed by Richard Trevithick in 1801. If that name sounds familiar ➤ **Fact check - only 0.4% of readers will remember him.** ✐ it's because he's the bloke who dug a disastrous tunnel under the Thames. Well, he wasn't quite as lousy at building trains. He'd heard about a guy called James Watt what had invented a steam engine. ➤ **Grammar check - 'who had'.** ✐ Sorry, he'd heard about a guy called James Who what had invented a steam engine. ➤ **Grammar check - *CF[8#MAS1&F}PBXc^{». ERROR.** ✐ Someone would shovel coal into a fire in the engine, this would heat up some water, the water would turn into steam, and the steam would push what's called a piston. A piston is a tube of metal that can move up and down, and which connects to the wheels and turns them round. Phew! And choo choo!

Thickitrev found a way of turning this engine into a mini train that could take passengers, and he called it the *Puffing Devil*. It puffed its way around Cornwall for a couple of days, at the amazingly slow speed of two miles per hour, which is as fast as my Great Aunt Prunella walks. And then he left it outside a pub while he ate a nice plate of roast goose and . . . the *Puffing Devil* exploded into smithereens. Oops. Just as well it didn't have any passengers. R-Trev didn't give up – he built a new, less flammable train called *Catch Me Who Can*. He put it on a circular track in London and charged people a shilling to ride on it, which is the equivalent of about a fiver these days. No one was really interested, unfortunately, so he decided that if nobody liked his lovely steam train, then he'd never make any more ever,

URIN SCORE
7/10
GOOD NAME,
ONE MARK
DEDUCTED FOR
STEALING FROM
PUFFING BILLY.

URIN SCORE
5/10
INTERESTING
NAME, BUT
REMINDS ME OF
A LEONARDO
DICAPRIO FILM.

HA HA!
HERE IT COMES!
HA HA! ANY
SECOND NOW!
HA HA!
PRETTY SOON!
JUST YOU WAIT!

I'M SOOOO
BORED.

ever, ever. Just like how some people – but never me – take their football home because other people keep scoring all the goals. ➤ **Fact check - you have done that five times this week already.** ✦ And that was the last Richard Trevithick ever had to do with trains. Luckily some other people got *on board* with them though. Get it? ➤ **Fact check - negative.** ✦

ROCKET MAN

George Stephenson was known as the 'Father of Railways'. He built a revolutionary new type of steam train called the *Rocket*, with the help of his son Robert, which is weird because I thought his son was called Railways. In 1829, everyone was going bananas for the cotton made in Manchester. Loads of cotton mills had just opened, and countries all over the world wanted it for their hoodies. ➤ **Fact check - the hoodie wasn't designed until 1934.** ✦ So the government built a nice long railway track from Manchester all the way to the coast at Liverpool. The only problem was, they didn't have any trains. They invited a load of engineers to bring along trains they had built, to see whose was strongest and fastest. It was like *Thomas the Tank Engine* meets *The Hunger Games*. George and Railways **Fact check - Robert.** brought along the *Rocket*, which could travel at a speedy (for the time) thirty miles per hour, and got ten out of ten from all the judges.

Soon, railways were being built all across the country. One of the main railway builders was a man called Isambard Kingdom Brunel, which is a coincidence because there was someone else called that who designed all those bridges. **Fact check - not a coincidence. It was the same person.** One of IsyKingyBroon's biggest achievements was building the Great Western Railway between London and Bristol. He even designed the station that the trains would use in London and named it Paddington, after his favourite bear. ➤ **Fact check – Paddington Bear is named after the station where he arrived from darkest Peru.** ➤

BITE THE BULLET

This chapter is like a big reunion of inventors from other chapters. We've already had Richard Trevithick and IKB, and now it's time to welcome back . . . Werner von Siemens! Not satisfied with building the first electric lift, he also designed the first electric train. I guess he just loved making things electric. And it's thanks to Werner that we can all whizz about the country so fast.

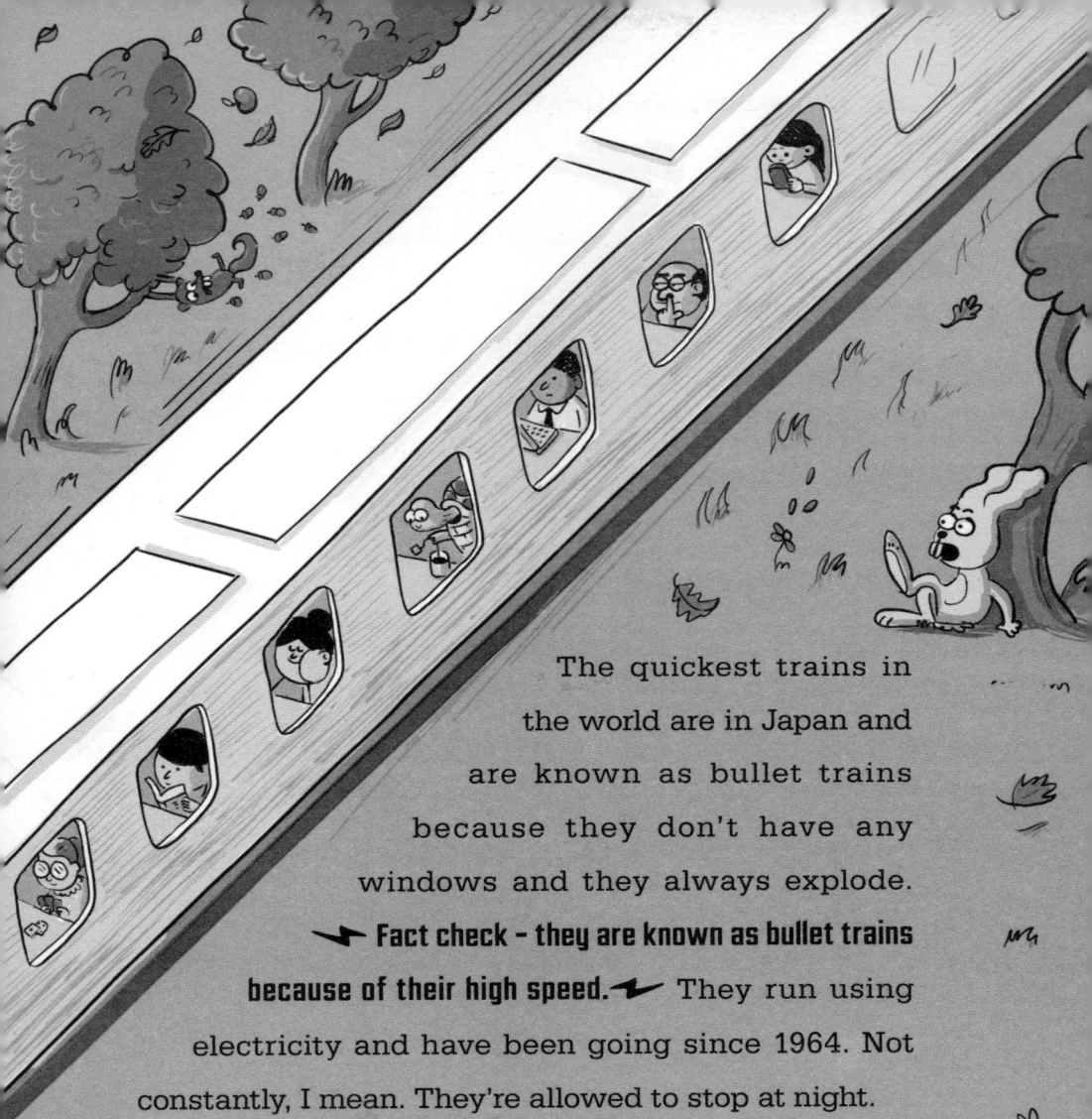

The quickest trains in the world are in Japan and are known as bullet trains because they don't have any windows and they always explode. ⚡ **Fact check - they are known as bullet trains because of their high speed.** ⚡ They run using electricity and have been going since 1964. Not constantly, I mean. They're allowed to stop at night.

The fastest bullet train in Japan travels at 375 miles per hour, which is probably as fast as a real bullet. ⚡ **Fact check - bullets travel at 1,700 miles per hour.** ⚡ OK, well, it's two-thirds as fast as a jumbo jet, and quite a lot faster than Richard Trevithick's rubbishy first effort.

ON YOUR BIKE

Say hi to Baron Karl Friedrich Christian Ludwig Freiherr Drais von Sauerbronn. 'Hi, Baron Karl Friedrich Christian Lu– actually, do you mind if we just call you Karl?'

BARON KARL FRIEDRICH CHRISTIAN LUDWIG FREIHERR –

DRAIS VON SAUERBRONN

Karl was a famous German inventor, which you could probably have guessed, because he's in a book about inventions. He created a device for printing piano music, the first meat grinder and a kind of slow cooker – so you know who to thank next time you're writing piano music while cooking mince. But in 1817, he also invented the bicycle, which you could probably have also guessed, because he's at the start of the section about bicycles. His design was pretty much like bikes are today . . .

except it was missing a couple of major things. See if you can spot what's missing from Karl's bike, which had:

- a wooden frame
- metal wheels
- a seat
- handlebars.

Yep, it didn't have any pedals or a chain, so cyclists would sit down and push themselves along with their feet, like Fred Flintstone in his car. ➤ **Fact check – only 3.2% of your readers will have watched the Flintstones cartoons.** ➤

Quick quiz – which three of these were some original names for the bicycle?

A) Dandy Horse
B) Mechanical Monkey
C) Swiftwalker
D) Velocipede
E) Spinning Arthur

If you've chosen A, C and D, you win the Empire State Building. (My lawyer, Nigel, has asked me to point out that I have no right to gift you the Empire State Building and, accordingly, you can be conferred no such ownership.) If you've chosen B or E, you have to hoover the Empire State Building.

PENNY DREAM WILL DO

Have you ever seen a picture or a film from the olden days, where some weirdigan with a long twirly moustache is riding a bicycle with an absolutely massive wheel at the front and a ridiculously tiny one at the back? In fact, it's still how my Great Aunt Prunella travels around. This bike was called the penny-farthing and it was popular about a hundred and fifty years ago.

It got this name because of its inventor, Penny Farthing. ➤ Fact check – it was named after two different-sized coins. ➤ OK, fine. The back wheel was like a tiny farthing (there were 960 farthings in a pound back then, for some strange reason) and the front wheel was like the much bigger penny (there were 240 pennies in a pound, for some even stranger reason).

The penny-farthing had rubber wheels instead of metal ones, but they were solid rubber rather than filled with air. That meant it was a slightly less bumpy ride than on Karl's bike, but you still wouldn't want to try eating a bowl of soup on one. ➤ **Fact check – inflatable rubber tyres were invented in 1887 by a vet called John Dunlop.** ➤

No one was particularly worried about sitting on a bike that was so high up, because people were used to riding horses everywhere. The only problem was that everyone kept falling off and smashing their heads open. In 1885, John Kemp Starley came up with an idea . . . why not make both wheels small? He even put springs in the saddle to stop everyone's bums getting so bruised. He called it the 'safety bicycle' and it looked pretty much like the bikes that we all ride today.

GO TO YOUR VROOM

The first-ever car was designed by Leonardo DiCaprio, an American actor famous for his roles in *Titanic* and – ➤ **Fact check – it was designed by Leonardo da Vinci.** ➤ Yes, that makes more sense, actually – I was wondering how he did that five hundred years ago. Leo DaVee was like that annoying person at school who's good at absolutely everything. (At my school it was me.) **Fact check – it was actually Charlie Davidson.** Leo was a brilliant artist and inventor, who came up with ideas for hundreds of things we still use today. Including the 'self-propelled cart'. It worked like a giant wind-up toy and had three wheels and no roof, so it was basically a big broken roller skate.

DA VINCI-EROO

The one quite major thing missing from Leo's idea was an engine, but this would have to wait until 1863, when a Belgian man called Étienne Lenoir invented a vehicle that he called the Hippomobile, because it had a big grille at the front that looked like a hippo's teeth. **Fact check – it was called the Hippomobile because 'hippo' means 'horse' in Ancient Greek, and the car was considered a kind of** metal horse. ⚡ The Hippomobile was basically a wooden wheelbarrow that you could jog faster than. But you could sit in it, and it had an engine and a steering wheel, so Étienne had invented the first proper car.

DRIVING US ROUND THE BENZ

You've probably heard of Mercedes-Benz. It's a fancy make of car – in fact, it's what I drive. ⚡ **Fact check – you drive a twenty-year-old van that you've stuck a Mercedes-Benz sticker on.** ⚡

Anyway, it was founded by a married couple called Carl and Bertha Benz. In 1886, Carl designed and built the Benz Patent-Motorwagen, which was the first car that people could actually buy. It included loads of incredible new features he'd invented that are still used today, such as the gear stick, the ignition and the radiator. The only problem was . . . no one wanted to buy his cars. There were two reasons for this. Firstly, not enough people knew about them. And secondly, they were pretty terrible. Luckily Bertha had a plan to fix both of these issues.

In 1888, Bertha B took her two sons and headed off on the world's first-ever road trip. She drove more than sixty miles across Germany, which was massively further than anyone had ever taken a car before. And, just like she'd planned, loads of people were interested in her amazing journey, and soon reports of Bertha's spangly new car were all over the newspapers and Instagram. ➤ **Fact check - not Instagram.** ➤ As she drove, she discovered a few issues with it. For example, it couldn't manage even the slightest hill. Bertha fixed this on the spot by adding an extra gear. Oh, and the brakes were totally useless. Not a problem for Bertha, who added a bit of leather to the brakes, casually inventing brake pads, which we still use today. By the time Bertha got back home, everyone knew about the Benz and wanted to buy one – and she and Carl were soon selling hundreds of cars a year.

CAN WE TURN THE RADIO OFF FOR A BIT?

HENRY AFFORDABLE

One major problem with Carl and Bertha's cars was that they cost more than a year's salary for the average family, so only really rich people could afford to drive them. A man called Henry Ford didn't think this was right – he reckoned everyone should be able to drive a car, not just the mega-posh. He started up a car company that is still around today called Mitsubishi ➤ **Fact check - it was called Ford.** ◄ and got to work on a bargain banger. The car he designed was named the Model T and, because of the clever way his factory worked, they were half the price of all the other cars available. His cars were made on an assembly line, which meant that every worker was responsible for adding the same part to every car that was made. So one person's job would be to add the door, then the next person along might add the handle, and the person after that would add the crumpled-up tissues and the old chocolate wrappers. ➤ **Fact check - ahem.** ◄ Because everyone was focused on a single job, it was a lot faster. In fact, it was so fast that two new cars could leave the factory every single minute, which was lucky because they ended up making fifteen million of them, and in 1918 half the cars in the United States were Model T Fords.

They weren't the most colourful cars in the world – as H-Fo said, 'You can have a car painted any colour that you want, as long as it's black.' Just like my books should say, 'You can have any kind of book you want, as long as it's amazing.' No fact check required, thank you.

But just because someone was a successful inventor, it doesn't mean they were a nice person. Henry Ford was extremely racist, and hated Jewish people in particular. Adolf Hitler – who started the Second World War and was probably the most evil person in history – called Ford his inspiration and kept a picture of him on his office wall.

The Model T Ford didn't have any fancy features like heated seats or a touchscreen display. It didn't even have a brake light or indicators – these were invented by the person with the best name in this whole book . . . a film star called Florence Lawrence.

It's time to use my robot butler's lie detector again to see if you know which one of these facts about Florence Lawrence is a complete Versailles.

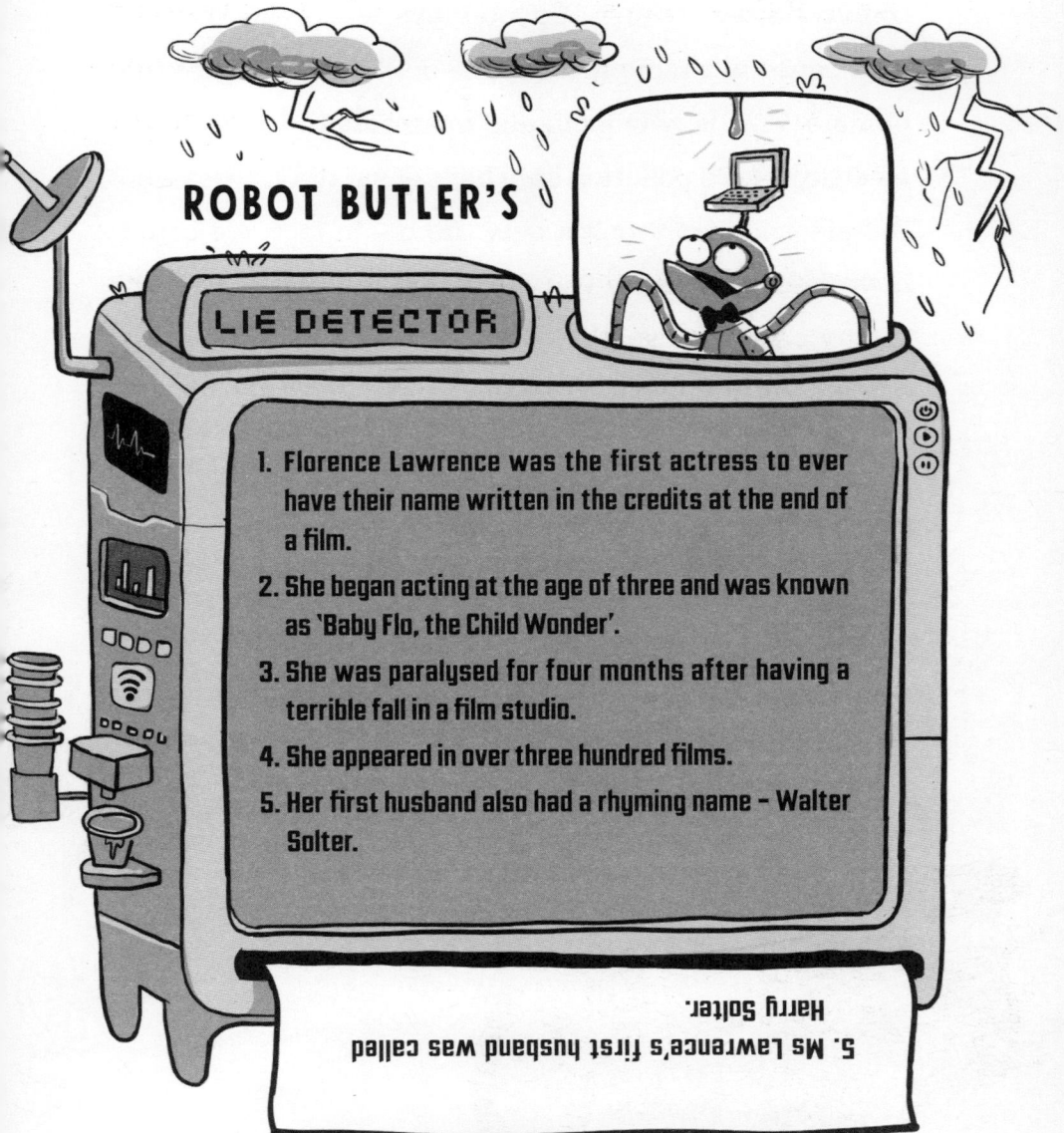

ROBOT BUTLER'S

LIE DETECTOR

1. Florence Lawrence was the first actress to ever have their name written in the credits at the end of a film.

2. She began acting at the age of three and was known as 'Baby Flo, the Child Wonder'.

3. She was paralysed for four months after having a terrible fall in a film studio.

4. She appeared in over three hundred films.

5. Her first husband also had a rhyming name – Walter Solter.

5. Ms Lawrence's first husband was called Harry Solter.

ELECTRIC AVENUE

Cars have absolutely changed the world, but unfortunately not just in good ways. The pollution that cars cause is a huge problem for our health, and for the climate. This is why more and more cars these days are electric, where pollution only happens if the driver farts. But electric cars aren't a new thing – in fact, we could have had them from the start instead of our belchy, fumey cars that smell worse than Pippin after she's eaten a bathful of baked beans.

While Carl and Bertha Benz were building their first cars, in a different part of Germany a man called Andreas Flocken was making an electric version called the

ELECTRIC VEHICLES TIMELINE

1890s
HUMMINGBIRD TAXI

1888
ELEKTROWAGON

Elektrowagen. And it worked! Loads of different makes of electric car started to appear, including an electric Porsche called the P1. There were even a bunch of electric taxis in London in the 1890s, which were known as hummingbirds because they hovered in mid-air. ➤

Fact check – because they made a quiet buzzing sound. ➤ We were really, really close to never having the whole pollution nightmare . . . but then people stopped buying them. The main problem was that they cost about three times more than petrol cars. About one hundred years after electric cars disappeared, people started getting worried about the environment and they made a big comeback. Hopefully this time they're here for good and the world will be saved. (Until the Octopus People of Zaarg get their tentacles all over us, obviously.)

2008
TESLA

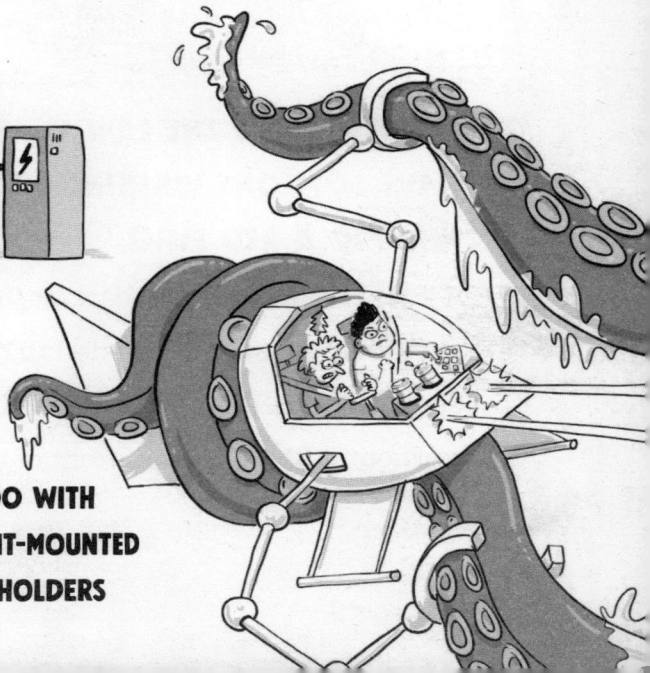

2075
ELECTRIC SKY-CAR 7000 WITH
ANTI-TENTACLE CLAWS, FRONT-MOUNTED
LASERS AND TWO DRINK HOLDERS

TRUE OR POO?

THE KING HAS HIS OWN SPECIAL TRAIN TO TRAVEL AROUND IN.

TRUE The Royal Train is basically a palace on wheels, and old Kingface sometimes uses it to travel around the country. It's bulletproof, to keep him safe, and has got a private bedroom, dining room and living room, as well as a bathroom with a full-sized bath – although the water probably sloshes out when the train drives round a bend.

IT USED TO BE THE LAW THAT EVERY CAR HAD TO HAVE SOMEONE WALKING IN FRONT OF IT, WAVING A RED FLAG.

TRUE When the first cars hit the roads, the government thought pedestrians needed to be warned that one was approaching. You could even be arrested for driving without a flag-waver in front of you.

IT TAKES A WHOLE MONTH TO CYCLE AROUND THE WORLD.

POO I'm afraid it takes a lot longer than that. The world record is held by Mark Beaumont, who cycled eighteen thousand miles in a bum-blistering two and a half months. It's probably quicker to take a plane, to be honest.

ADAM'S ANSWERS

WHEN WAS THE FIRST-EVER SPEEDING TICKET?

In 1896, a man called Walter Arnold was arrested for driving through a village in Kent at four times the speed limit! This is slightly less shocking when you hear that he was driving at eight miles per hour and the speed limit was two miles per hour. I'm surprised that snails didn't get arrested for breaking the speed limit back then too.

WHO INVENTED THE AIRBAG?

Airbags were invented in 1921 by a couple of dentists called Harold Round and Arthur Parrott who got bored of teeth – fair enough – and decided to make driving

URIN SCORE
2/10
AIRBAG IS AN INACCURATE NAME – IT SHOULD BE CALLED NITROGEN BAG.

safer. But it took another fifty years before airbags actually started to appear in cars. If a car detects that you're in a crash, then it quickly inflates a load of bags at the front and side of the car with nitrogen. This means that instead of a driver's head hitting the steering wheel, it hits a big pillow and hopefully prevents any serious injuries.

HOW MANY BIKES ARE THERE IN THE WORLD?

Oh, just a billion of them. Half of them are in China and more than one hundred million are in the United States. There are about twenty million in the UK, which is about one for every three people – although I'm not sure how three people get on a bike together.

EASY!

DREAMVENTIONS

I've come up with a lot of great ideas in my dreams. A sofa made out of marshmallows, for example. Extremely fluffy, very comfortable, with the tiny minuscule downside of the fifty-two million wasps who ended up living in my house. Then there were stork forks (cutlery for birds) and spring onions (springs made out of onions).

But have any of them changed the world? Well, stork forks were pretty popular. ⟶ **Fact check – only eight stork forks were sold, and six of those were to members of your own family.** ⟵ Here are some other inventors who came up with incredible ideas while they were snoozing.

DNA

DNA (which is short for deoxyribonucleic acid – see if your parents know that – if not, tell them to go back to brain school) is what makes you . . . you! The colour of your hair, the shape of your ears, the smell of your farts – DNA is like the code that programs everything about you. We know all about it thanks to four brainboxes called Rosalind Franklin, James Watson, Francis Crick and Maurice Wilkins. One of the most interesting things they discovered about DNA was its shape: two spirals weaving in and out and around each other. And how did they work this out? Well, James Watson was having a nap and had a dream about a funny double-sided staircase, then doodled it down and – hey presto! DNA!

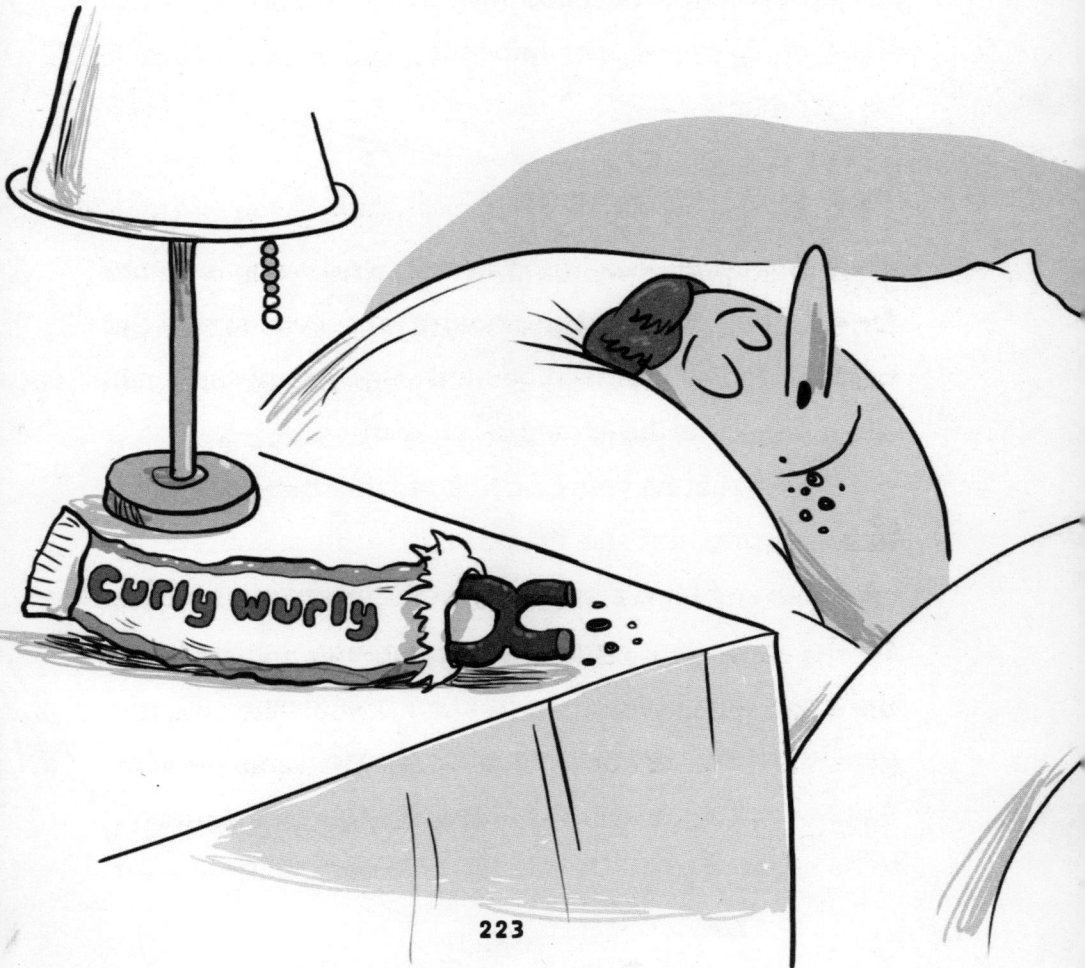

THE PERIODIC TABLE

Dmitri Mendeleev had spent ten years trying to work out a way of putting all the chemical elements into some kind of order. A periodic chair? No. A periodic carpet? No, that's not right either. Then when he was sleeping one night the answer just appeared to him. A periodic table! It was an extremely useful discovery, which changed the course of chemistry forever, but, I'm not going to lie, that's a pretty boring dream, Dmitri. Most people dream about school getting cancelled or flying baboons.

THE SEWING MACHINE

Elias Howe had been trying to design a sewing machine for ages and ages, but he couldn't get the shape of the needle right, so his machine didn't really work very well. Then one night he was zzz-ing away when he had a nightmare that he was on a holiday that had gone quite badly wrong, and he was about to be executed. He was being carried away by some warriors holding spears when he noticed that the spears had holes in the top . . . He woke up excited for two reasons. Firstly, those spears would be the perfect design for his needles. And secondly, it was just a dream and he wasn't going to be speared to death by a load of angry warriors.

FRANKENSTEIN

When I have a nightmare, I wake up in a cold sweat and can only calm down by calling Pippin upstairs for a cuddle. And then I immediately regret it because she smells of dog food and fox poo. When Mary Shelley had a nightmare two hundred years ago, she went to her desk and wrote it all down – she'd just dreamed up the story of *Frankenstein*.

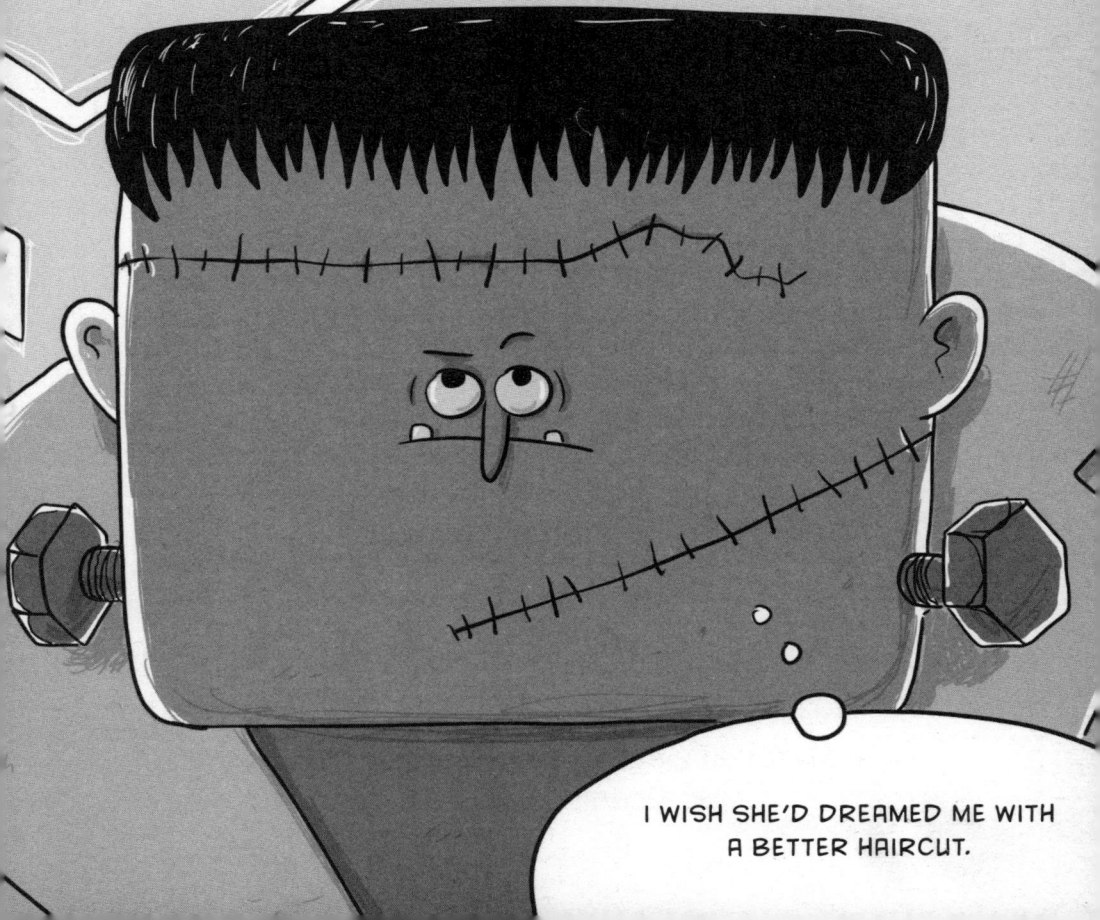

I WISH SHE'D DREAMED ME WITH A BETTER HAIRCUT.

TRAVEL: SEA AND AIR

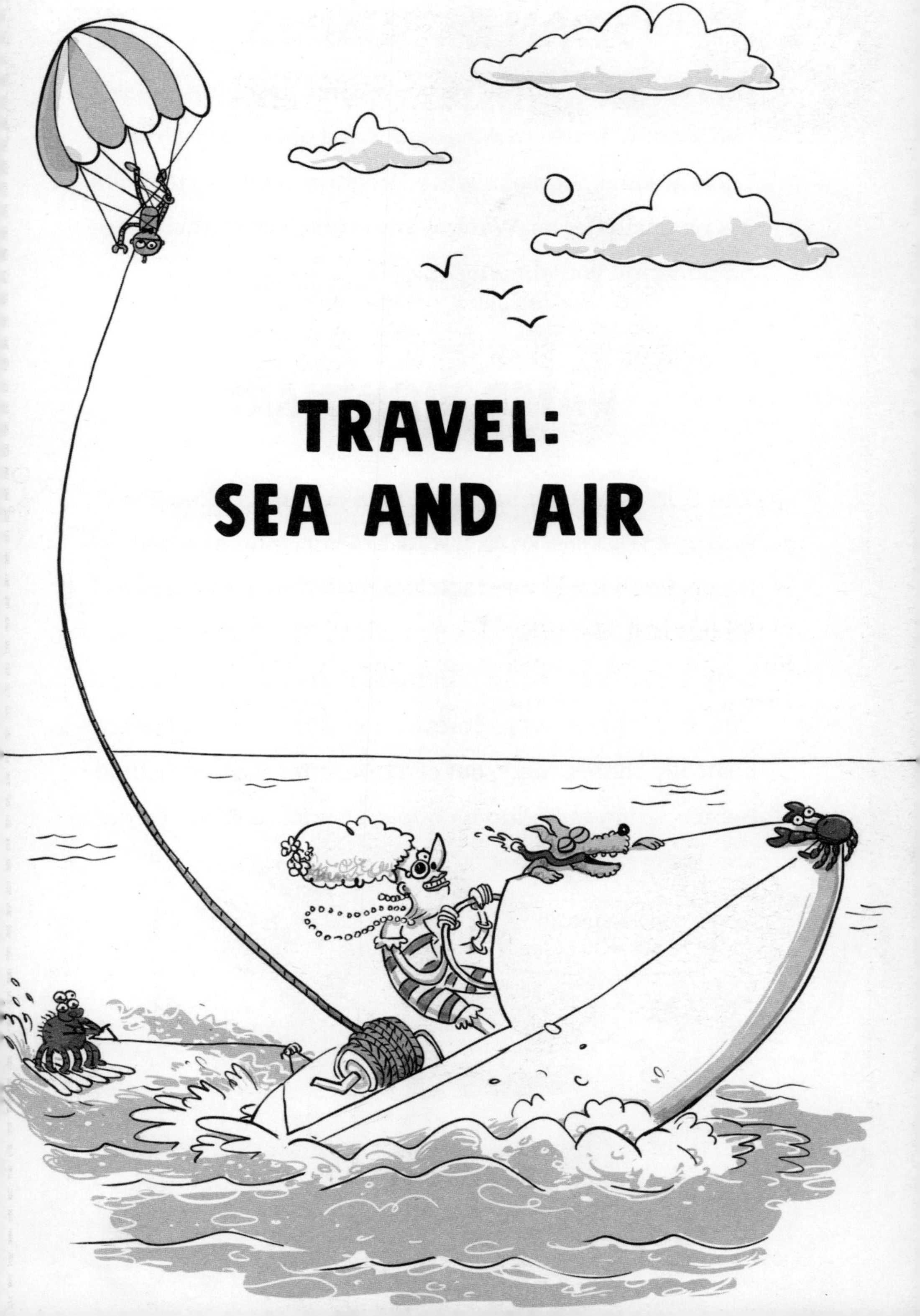

Cars and trains are all very well and good, but what if you want to travel to America? You'd get extremely wet. Luckily some people invented boats and planes to make it a bit easier for us. Want to know who? Well, tough, I'm going to tell you anyway.

WATER NIGHTMARE

The first boat was invented by a woman called Flo T. Raft. ⚡ **Fact check - no, it wasn't.** ⚡ It was invented by Ro-Ro-Rosie Boat? ⚡ **Fact check - no one knows who invented the first boat.** ⚡ What we do know is that humans have been making boats for hundreds of thousands of years. The first boats were known as 'dugouts' and were basically canoes made out of a tree, after the middle had been . . . dug out. Add a couple of oars, and you've got yourself a rowing boat.

HEY! THIS IS MY HOUSE!

SAIL LA VIE

The next big idea came from the Ancient Egyptians about six thousand years ago, when they realized that if you stick up a nice tall mast and hang a sail off it, the wind will do all the hard work and blow you along. I'm not sure if they discovered this deliberately, or if someone was just drying their underpants on a boat. Boats back then were made out of reeds, which is a type of tall grass. I know, this sounds like a terrible thing to make a boat out of, but if you tied the reeds up really tightly into bundles it pretty much worked. I say 'pretty much worked' because it often let loads of water inside.

This was a problem that was bothering an Ancient Greek mathematician called Archie Medes. **Fact check – he was called Archimedes.** He had built an absolutely enormous ship called the *Syracusia*, which was three storeys high, as long as a football pitch and could hold two thousand passengers. It was extremely luxurious and featured a temple, a library, a gym and a drive-through McDonald's. **Fact check – the first McDonald's restaurant was built two thousand years later, in 1955.** Now, if you've built the biggest and best boat in the world, then you definitely don't want it to sink, so Archie invented a way of sucking up any water that had got inside and squirting it back into the sea. He called it the Archimedes Screw. His gadget was a screw inside a tube and, if you turned it, it would lift water from the bottom to the top. And we still use it today – cheers, Arch! Here's a diagram of the Archimedes Screw or, if you don't care, here's a picture of Winnie-the-Pooh eating some glue.

URIN SCORE
4/10
VERY
BOASTFUL.

Fact check – you have named four books after yourself.

(My lawyer, Nigel, has asked me to point out that you shouldn't eat glue, even if you're a cartoon bear.)

Archimedes's other discoveries could fill a book themselves, but they included how to work out the area of a circle (πr^2) and a sphere ($4\pi r^2$), and the odometer, which is a device that measures how bad your breath is. ⚡**Fact check – it measures distance.**⚡ He was also really good at designing levers and pulleys, which meant people could move extremely heavy objects without having biceps the size of a fridge. He once said, 'Give me a place to stand and I shall move the world.'

It's time to once again activate my robot butler's lie detector, so you can work out which of these facts about Archimedes is a complete chicken thigh.

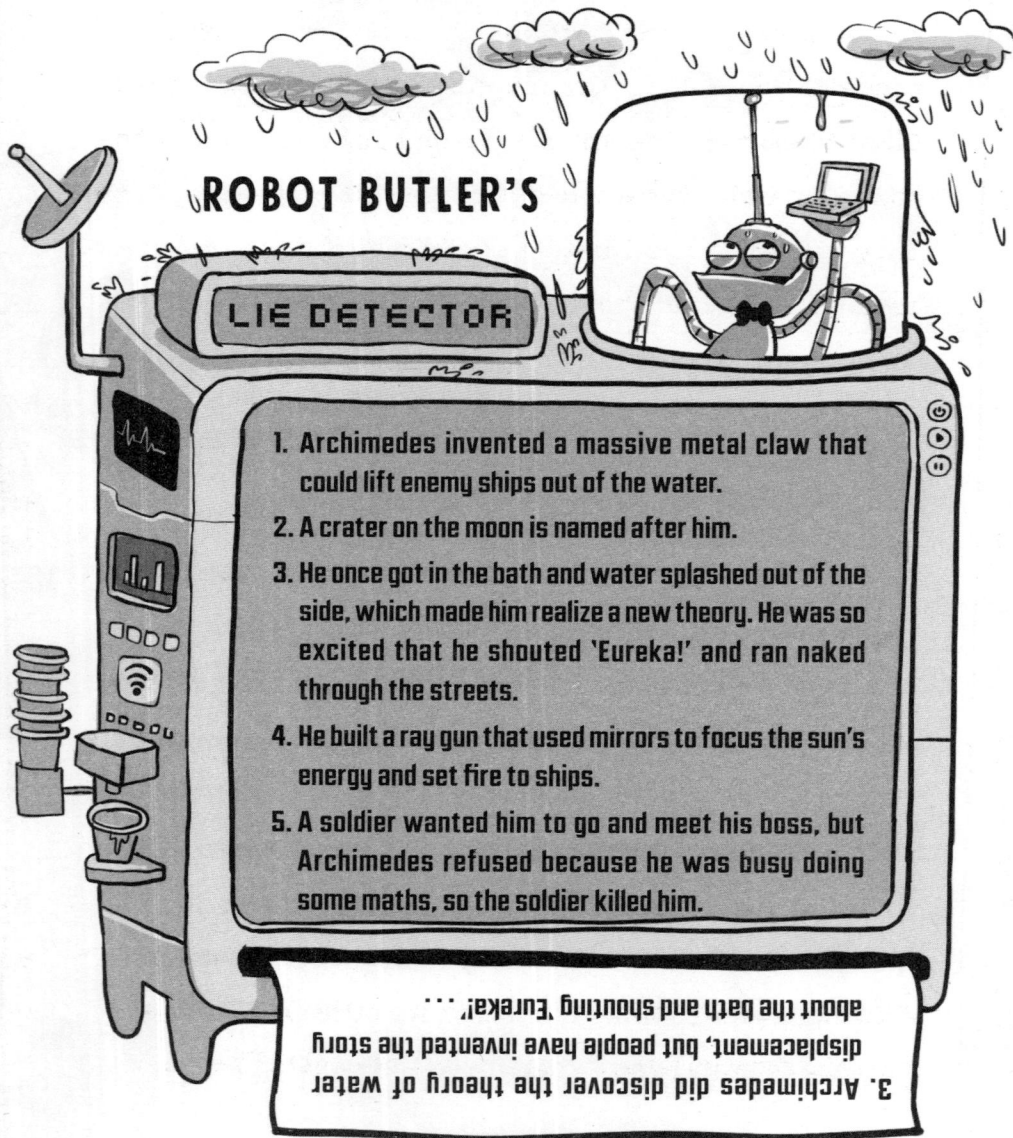

ROBOT BUTLER'S

LIE DETECTOR

1. Archimedes invented a massive metal claw that could lift enemy ships out of the water.

2. A crater on the moon is named after him.

3. He once got in the bath and water splashed out of the side, which made him realize a new theory. He was so excited that he shouted 'Eureka!' and ran naked through the streets.

4. He built a ray gun that used mirrors to focus the sun's energy and set fire to ships.

5. A soldier wanted him to go and meet his boss, but Archimedes refused because he was busy doing some maths, so the soldier killed him.

3. Archimedes did discover the theory of water displacement, but people have invented the story about the bath and shouting 'Eureka!' . . .

STEAMING AHEAD

Ships were blown by the wind for thousands of years until the 1800s, when people tried using that newfangled piece of technology, the teleporter. **Fact check - the steam engine.** The first steamships used their engines to turn massive great wheels on the side of the boat, but then someone tried putting an Archimedes Screw in reverse and found that it worked as a propeller – these are still used today.

One of the main steamship designers was a man called . . . Isambard Kingdom Brunel. This guy made everything – when did he sleep? The first ship he built, the SS *Great Western*, was the largest in the world at the time and could travel from Bristol to New York in two weeks, which was twice as fast as the rubbish old ships with sails. Soon, even bigger and far more luxurious ships were being built to take passengers to America. They became extremely popular, and people would queue up to travel on ships called things like the *Teutonic*, the *Majestic*, the *Olympic* and . . . umm . . . the *Titanic*. They got slightly less popular for a while after the *Titanic*.

DEEP DIVE

It's quite possible that Leonardo da Vinci was a time-traveller. Even though he lived five hundred years ago, his notebooks contain drawings of inventions that people didn't make until really recently, from tanks to air conditioning. He even drew things that don't exist yet, like flying bikes. And one of his amazing inventions (or cheating bits of time travel) was the submarine. He called it the 'ship to sink another ship', and it was designed to sneak up on enemy boats and never get seen. Unfortunately, Leo's submarine didn't get made – or maybe it did and he designed it so well that nobody has ever seen it.

PLEASE IGNORE ME

The first person to build a submarine that worked and that you could steer around was a Dutch engineer called Cornelius Drebbel, in 1620. He made it out of wood, then covered it with leather and slathered it with grease, so it wouldn't leak – which is never great if you're on a submarine. His system worked, and there were no dribbles for Drebbel. It was powered by people rowing it, and could carry sixteen passengers, who would get their air from tubes that went up to the surface, like enormous snorkels. People who went for a ride in the Drebbelmobile included King James I. Which is probably short for King James the First King to Go in a Submarine.

⚡ Fact check – it is not. ⚡

I'M JUST A NORMAL TROUT

Submarines these days are slightly more advanced, with no snorkels involved. Because they swim around in total darkness, without any windows, they know how to avoid oil rigs, boats, underwater mountains and mermaids ➤ **Fact check – not mermaids.** ➤ thanks to sonar. Sonar was yet another thing that was invented by the most visionary inventor of all time. That's right – me. ➤ **Fact check – Leonardo da Vinci.** ➤ It works by letting out a very high squeak of sound into the water and analysing the sound waves that bounce back. This is also how dolphins know what's going on around them, and it clearly works because you never see a dolphin with a plaster on their nose from bashing into a wall.

AM I HOVVERED?

The hovercraft was invented in the 1950s by a cockerel called Sir Christopher Man. ➤ **Fact check – a man called Sir Christopher Cockerell.** ➤ The hovercraft is a boat that's pushed forward by big propellers on its back, and it sits on a pillow full of air, which means it can travel across any surface, whether it's land, water, ice or poo. Sir Chris made his model of the first hovercraft out of a vacuum cleaner and a couple of tins of cat food, which definitely stank but proved that a boat could float just above the water. Hovercrafts could travel extremely fast – over eighty miles per hour, which is quicker than cars are allowed to go on the motorway. They became very popular as a way of crossing the English Channel. You might think that travelling on a cushion of air would be smooth and relaxing, but it was about as bumpy as using a pneumatic drill while riding a pogo stick on a cobbled street. Hovercrafts became a lot less popular when the Channel Tunnel opened, because that meant you could get to France and drink a hot chocolate without it ending up all over your hair, face and trousers.

PLANE AND SIMPLE

Ever since humans first looked up and saw a bird swooping overhead, we've been desperate to fly. Why should pterodactyls have all the fun? ➤ **Fact check – pterodactyls became extinct millions of years before humans were on Earth.** ✦

But it wasn't until two hundred and fifty years ago that anyone successfully made it into the air. And, more importantly, managed to stay up there. Joseph-Michel and Jacques-Étienne Montgolfier were brothers who lived in France in the 1700s. French people really love a double-barrelled first name, don't they? Jo-Mic was staring into the fireplace one evening when he saw sparks flying up into the chimney and wondered if this might be a way to finally get into the sky. Before long, he and his brother had stitched together the world's first-ever hot-air balloon and were sending their first passengers into the sky in a balloon they'd made out of a worryingly flammable mixture of paper and cloth.

Well, I say passengers – they were a duck, a sheep and a cockerel. The duck and the cockerel can't have been too worried – if anything went wrong, they could just fly away – but the sheep must have been absolutely pooing the basket. But the sheep didn't have anything to worry about, though, as the balloon landed safely about ten minutes later.

The Balloon Bros thought there was some magical flying vapour being produced by the smoke in the fire, which they called Montgolfier gas. It turns out there wasn't any special gas at all; it was just plain old air. When air gets hot, it gets lighter, which is why you put smoke detectors on the ceiling, not on the floor. And that's why their balloon floated. There's only one teeny-weeny problem with hot-air balloons – you can't steer them; you can only make them go up and down by adjusting the flame. If you want to go west, then you need to find a gust of wind that's going west – a bit like sky-hitchhiking.

TRY TO FOCUS ON YOUR BLEATING.

INCY WINCY GLIDER

It would be quite annoying if you thought you were getting on a flight to Spain but you actually arrived in Bulgaria because of where the wind blew you, so the next step was building a plane that you could actually steer. If you've ever seen any big birds swooping around the sky without moving their wings, that's known as gliding. ➤ **Fact check - Big Bird is a character from *Sesame Street* who cannot fly.** ➤ I meant big types of birds, like eagles. Because nobody had invented the engine yet, the very first planes had to glide like this, and they were called gliders.

URIN SCORE
5/10
FAR TOO
OBVIOUS.

George Cayley lived in an absolutely massive house. Lucky old George. Back in 1804, he would go up to the top of his absolutely massive staircase and throw different types of wooden wings down to his absolutely massive hallway. A whole bunch of smashed-up wings, tiles and bannisters later, Georgie realized that the way to make a wing work is for it to be flat underneath and curved on top. Time for a mini science bit. Soz.

MINI SCIENCE BIT

Because air has to travel a bigger distance to get over the curved part of the wing on top, it has to go faster. The faster that air moves, the lower the pressure it has, which means the wing has higher pressure below it than above it. Things move from high pressure to low pressure, so this means that the wing moves upwards . . . so it flies. Here's a diagram explaining it, plus an alternative picture of a mosquito eating a burrito in case you couldn't care less.

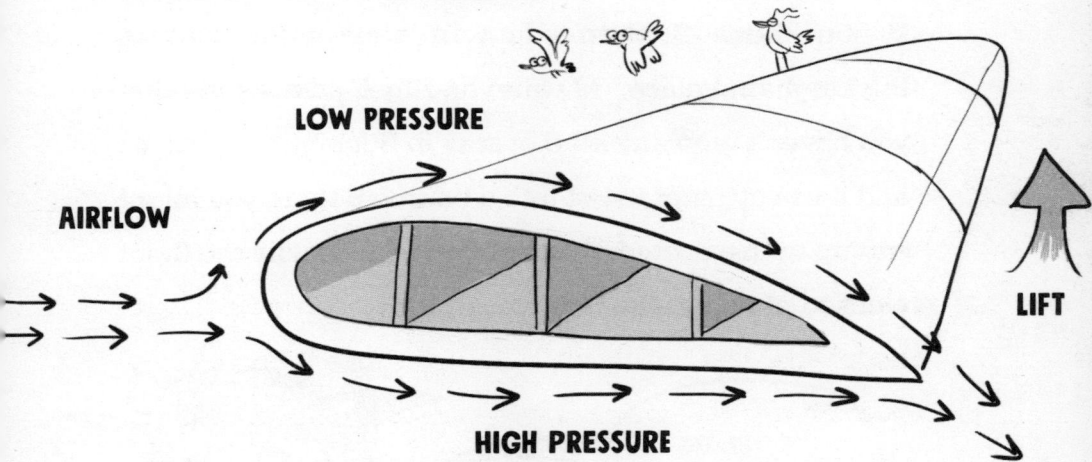

LOW PRESSURE

AIRFLOW

LIFT

HIGH PRESSURE

In 1848, Georgio stopped chucking wings down the stairs and finally built a glider. But who took its first-ever flight?

A) George Cayley himself
B) His dog, Bertram
C) A random ten-year-old boy
D) The prime minister, Lord Russell
E) My Great Aunt Prunella

If you chose C, then you win a six-week stay in Buckingham Palace. (My lawyer, Nigel, advises me that you haven't won any kind of stay in Buckingham Palace and if you attempt to get into a bedroom there you might end up with a six-month stay in prison.) Luckily the flight went to plan and the boy was fine.

WRIGHT PLACE, WRIGHT TIME

Brothers Orville and Wilbur Wright were probably the most important people in the history of flying, unless you count Samuel Aeroplane. **⤚ Fact check - my biography module can find no record of Samuel Aeroplane.⤜** Instead of going to university, Orv and Wilb opened a bike shop and used the money from that to pay for their true love – inventing perfume that smelled like farts. **⤚ Fact check - inventing aeroplanes.⤜**

They first tried out their new invention, the *Wright Flyer*, in 1903. It didn't look much like the planes we might go on holiday in – the wings were basically two rows of lolly sticks with supports in between. But it worked! Well, a bit. It flew for twelve seconds, across the spookily named Kill Devil Hills at about the same speed you can run.

URIN SCORE
2/10
THEY CLEARLY ONLY THOUGHT ABOUT THIS NAME FOR TWELVE SECONDS.

TEA?
COFFEE?
PRETZELS?

URIN SCORE
0/10
EVEN WORSE.
THE WORST
NAME I'VE EVER
HEARD.

They tried it out a few times until a gust of wind tipped it upside down and it smashed into pieces. Their next attempt, the *Flyer 2*, wasn't much better but the one after that, the *Flyer 3*, was far more successful. In 1905, Wilb managed to fly for nearly forty minutes in the *Flyer 3* – he might have gone on longer, but he forgot to fill up the petrol tank before he left.

URIN SCORE
-1/10
NOPE, THEY
MANAGED
TO MAKE THE
NAME WORSE.

Even though they'd invented the first-ever working plane, only one person in the country was interested enough to write about them. He was called Amos Root and he ran a magazine about bees. Amos watched them fly (the brothers, not the bees – he saw bees flying all the time) and then wrote about it, but not many people were reading bee magazines, so the story didn't generate much buzz. **Fact check - my joke-assessment module informs me this has a 6% humour level.** Other journalists just didn't believe that the Wright brothers had even done it – the *New York Herald Tribune* asked if they were 'flyers or liars'.

I'm sure you would never do this, but you might be aware of some other people (definitely not you, absolutely not) who storm out of the room when they're upset. Well, the Wrights sort of did this. They left the USA and went off to France, home of the hot-air balloon and double-barrelled first names, to see if the French would be more *intéressés*. And they were! The French were amazed as Wilb flew his plane in circles and figures of eight and spelled out the words 'TOLD YOU, AMERICA' in twelve different colours of smoke. **Fact check - the smoke writing did not occur.** Soon, the Wrights were selling their planes everywhere – well, mostly to the army, because there weren't any airports yet. They even let the USA buy some.

They had also accidentally invented a brand-new form of entertainment – the air show! Up to half a million people would come along and watch daredevil pilots loop the loop in their incredible (and incredibly rickety) new planes. That's more than five times the number of people that can fit into Wembley Stadium.

THAT'S CHEATING.

One of the most famous air-show pilots was Bessie Coleman, known as Brave Bessie, for obvious reasons. She was the first Black person to get an international pilot's licence, and performed the most amazing tricks in her planes. Sadly, her story ended in tragedy in 1926 when a faulty plane she was a passenger in crashed to the ground, but she is still remembered as one of the greatest pilots of all time.

NEXT FLIGHT DEPARTING

The first flight that you could buy a ticket on was in 1914, going from St Petersburg in Florida all the way to Tampa in, umm, Florida. Because there still weren't any airports yet, it had to take off and land on water, which sounds totally terrifying and I'm glad that they've had the good sense to build some runways now. Planes became really important in the First and Second World Wars, and America alone built over three hundred thousand planes. Then, after the Second World War, some of these ended up being bought by airlines to carry passengers.

It wasn't particularly fun travelling by plane in those days. They were extremely noisy, cold and bumpy. In fact, so many people got sick that lots of planes had nurses on board to look after the passengers. Plus, the planes would have to stop all the time to refuel. And they crashed quite a lot. Oh, and tickets were incredibly expensive.

Things improved a lot for passengers when planes swapped their propellers for much more powerful jet engines, which meant that people could finally fly to America. Another big improvement was pressurized cabins (posh air conditioning to make sure the air is nice and breathable even when you're really high up). And because planes could now fly above all the turbulence, there wouldn't be any more vomit flowing down the aisles.

In 1976, a plane called the Concorde was built with four mega-powerful jet engines and super-streamlined wings. It could fly at twice the speed of sound and get from London to New York in only three and half hours, which is how long it takes to get a train from London to Plymouth (which doesn't have as many skyscrapers). It was extremely expensive making the Concorde fly so the company kept losing money, and its last flight was in 2003. You can still take the train to Plymouth though.

URIN SCORE
9/10
GREAT NAME, BECAUSE IT WAS MADE JOINTLY BY BRITAIN AND FRANCE, AND 'CONCORDE' MEANS 'IN AGREEMENT' IN BOTH LANGUAGES.

WHAT GOES UP MUST COME DOWN

Maybe this chapter should have been called 'Stuff Leonardo da Vinci Came Up With Hundreds of Years Ago'. Not content with inventing cars, submarines and bubblegum ➤ **Fact check – bubblegum was invented in 1928 by an accountant called Walter Diemer.** ◄ Leo also doodled a design for a parachute. It was a big wooden square with

a pyramid of cloth on top, which a person could dangle from underneath. LdV never actually built one because he was so unbelievably lazy, but about twenty years ago a British skydiver used those original designs to make one, and . . . it worked! If I was that skydiver, I'd have probably worn a proper parachute as well, just in case it didn't . . .

The first person to safely fly to earth using a parachute was Louis-Sébastien Lenormand (you guessed it, he was French), who managed it in 1783. He also invented the word 'parachute', which means 'stop falling'. Even though it worked, it wasn't very practical – it was the size of two beds, so not the kind of thing you could strap on in a hurry.

In 1910, Katharina Paulus was getting worried every time her husband went off to work. They were both daredevil performers, but his favourite trick was to jump out of hot-air balloons and parachute down to earth. I feel just as nervous when my husband goes off to his job as a dentist for tigers. ➤ Fact check – your husband actually makes badly reviewed television shows. ⚡ So Katharina designed a much better parachute that could fit into a backpack for the first time and then open safely when it was needed. Unfortunately, one day it didn't open and he splatted to the ground . . . oops.

ARE YOU SURE THAT'S THE RIGHT BACKPACK?

ONE SMALL ZEP FOR MAN

While Orville and Wilbur Wright were busy doing their thing, over in Germany Count Ferdinand von Zeppelin had a very different idea about how we should all be flying around. In massive helium balloons. The Zeppelin was shaped like a torpedo, longer than a football pitch, and made of an aluminium frame covered in cloth. It could fly because it contained loads of enormous balloons full of a gas, such as hydrogen, that's lighter than air.

Slightly disgustingly, these balloons were made using cows' intestines. A quarter of a million cows needed to donate their intestines for every single Zeppelin that flew. In fact, so many were needed that for years sausages were banned in Germany, so that there were enough intestines to make Zeppelins, instead of using

them all for sausage skins. Zeppelins really took off **➤ Fact check – my joke-assessment module informs me this has an undetectable humour level.➤** and were used for everything from dropping bombs in the First World War to trips across the Atlantic.

This was until 1937, when the *Hindenburg* disaster happened. As you can probably guess from its name, the *Hindenburg* disaster wasn't a good thing. The *Hindenburg* was a huge Zeppelin that had flown from Germany and was travelling across America. Even though hydrogen is lighter than air, so it's good for making things float, it's also extremely flammable, and this meant that when there was a tiny spark of electricity on board, the whole thing exploded, killing thirty-six people. Today there are still some Zeppelins flying around, but you'll be pleased to hear they don't use flammable gases like hydrogen any more.

HELLOCOPTER!

In case you were worried that Leonardo da Vinci hadn't invented anything for a few paragraphs, here he is again. He drew a picture of the 'aerial screw', which was the same basic idea as the helicopter: a big spinny sail that had to be twisted round by a bunch of people standing on a platform below it, until it rose into the air. He never actually built it – classic LdV – in fact, nobody made one until about a hundred years ago, when *yet another* set of brothers, called Louis and Jacques Bréguet, made the Gyroplane.

The Gyroplane wasn't an amazing helicopter, to be honest. It could only fly up to knee height, you couldn't steer it, and four people needed to stand on the ground and hold it to stop it from crashing. But the Brég Bros had managed to get a helicopter into the air for the first time.

Helicopters weren't available in the shops until 1939. Igor Sikorsky had been a massive Leonardo da Vinci fanboy ever since he was a young child. A bit like how kids today are totally obsessed with my books.

✈ **Fact check – 87% of your readers described your books as 'OK, I guess'.** ✈ Anyway, Igor spent years and years trying to make a helicopter work, and eventually came up with the idea of putting a small rotor on its tail as well as a big one on the top. When the rotor on top spins, it pushes the air downwards, and the helicopter lifts up. The rotor on the tail helps to keep it balanced, and allows the pilot to steer it wherever it needs to go. This is how helicopters still work today. Helicopters can be used for lots of things that planes can't, such as rescuing people from mountains and bad accidents. This is because they can fly very low to the ground, land in small spaces, move in any direction, hover, and even fly upside down, which is obviously a massive 'no thanks' from me.

JUST BECAUSE YOU CAN DOESN'T MEAN YOU SHOULD!

TRUE OR POO?

THE FIRST JETPACK WAS BUILT IN 1419.

POO It was quite a lot later – the first person to design a rucksack with a jet engine inside was Alexander Andreev in Russia in 1919. The first working jetpack wasn't actually made until 1961. It was called the Rocket Belt, it was ten times louder than a chainsaw, and it only flew for twenty-one seconds. Not quite Iron Man. But the technology has got much better and soon paramedics will be using them to rescue people on mountains where even helicopters can't reach.

BLACK BOXES ON AN AEROPLANE ARE BRIGHT ORANGE.

TRUE All planes need to have a 'black box', or flight data recorder, which keeps a record of everything that happens on every journey, in case the plane has an accident. (I mean in case it crashes, not in case it wees itself.) But the black boxes aren't black at all – they're painted orange to make them much easier to find if there is a crash.

IT'S CONSIDERED BAD LUCK FOR ANYONE TO WHISTLE ON A ROYAL NAVY SHIP.

POO But almost true. Superstitious sailors used to think that whistling on a ship could attract a storm, and even now people don't like it one little bit. But one member of the crew is allowed to whistle – in fact, they're encouraged to. The ship's cook can still whistle away, because it proves that they're not eating the food as they work.

ADAM'S ANSWERS

WHAT HAPPENS IF A HELICOPTER'S ENGINE STOPS WORKING?

It's not ideal, but it's not a total absolute complete disaster. If a plane's engines fail, then it basically becomes a big glider, and the pilot can hopefully get it safely down to the ground. Helicopters don't have nice big wings to glide on, but don't panic – if their engines stop, they don't just crash to earth like an asteroid. You might have noticed that certain trees, such as the maple and the sycamore, have funny V-shaped seeds that spin very slowly down to the ground – they're often called helicopter seeds. This is what happens if a helicopter's engine stops: the force of the air makes the rotor spin like a sycamore seed, and it goes down to earth a lot slower than you might think! But, I'll be honest, it's probably best if the engine keeps working for the whole journey.

WHAT'S THE MOST EXPENSIVE BOAT IN THE WORLD?

What's the point of being a billionaire if you can't spend the summer on a ridiculous yacht? That's certainly what a Russian businessman thought when he bought a boat called the *Dilbar* for about half a billion pounds. It's got enough bedrooms for 132 people to stay in, a helipad and an absolutely enormous swimming pool. If you're thinking of buying it, I should point out that every time you fill up the tank it costs half a million pounds, so make sure you get enough pocket money for that.

WHAT'S THE LONGEST TRIP EVER TAKEN IN A HOT-AIR BALLOON?

The entire way around the world! Remember Bertrand Piccard, the man behind the solar plane?

⚡ Fact check – 0.0002% of readers remember. ⚡ Well, in 1999, he spent twenty days taking a 25,000-mile trip around the globe in a hot-air balloon. But, like I said, you can't steer a hot-air balloon, so maybe he just meant to go to Sainsbury's and see the marvellous Bread Section.

SORRY, WE'VE ONLY GOT A TESCO EXPRESS.

FREEVENTION$

Not all inventors get super rich. For example, even though I invented the inflatable lawnmower, I still have to write books for losers to read so I can afford to pay my bills. Sorry, I didn't mean 'losers', I meant highly intelligent, wonderful children. ➤ **Fact check – lie detected.** ➤ Here are inventions by a few people who should have become zillionaires . . . but didn't.

INSULIN

Diabetes is a common condition where someone's body isn't good at making insulin, which affects their sugar levels. If a person doesn't produce any insulin at all, then it's important that it gets replaced, either with insulin injections or using a little pump. The scientists who discovered how to replace insulin were called Frederick Banting and Charles Best, and they realized that their discovery was going to save the lives of millions of people all around the world. They didn't want to get rich from it – instead, they wanted anyone who needed insulin to be able to afford it, so they gave the patent (the right to make it) to the world for free. Thanks, Frederick and Charles – you're the banting! Hang on . . . I mean, you're the best!

THE SAFETY PIN

Walter Hunt was always inventing things – a knife sharpener, a road-sweeping gadget, a machine for making nails. But his biggest and best invention came in 1849: the safety pin – billions of them were soon being produced every single year. Unfortunately, he sold the patent to it for a few hundred pounds, and that's all the cash he ever made from it. But Walter didn't mind – he just moved straight on to his next invention: shoes with suction cups underneath, so that acrobats could walk up walls. I have to say that I don't think he sold quite so many of those . . .

SUCTION SHOES
SALE

£50 £40 £30 £20
JUST TAKE THEM!

KARAOKE MACHINES

It is a fact that I have the most beautiful singing voice in the world. ⚡ **Fact check – further lie detected.** ⚡ After having dinner with my friends, they like nothing more than going to a karaoke bar so they can listen to my perfect renditions of pop songs. **➤ Fact check – your friend Bruce has announced that 'I would rather grate off my own nose than hear a second more of your terrible singing'.➤** I can only perform such masterpieces because of a man called Daisuke Inoue, who invented the karaoke machine in 1971. While karaoke has brought joy to the hearts of anyone who has heard me sing, it sadly didn't bring any cash to Daisuke, who didn't take out a patent for his invention. These days, there are over one hundred thousand karaoke bars in China alone.

MONEY, MONEY, MONEY, MUST BE FUNNY . . .

FIDGET SPINNERS

Not quite as important to the world as insulin, but more fun to play with, are fidget spinners – little spinny bits of plastic that you put on your finger and whizz round.

An engineer called Catherine Hettinger had the patent for a design very similar to this about thirty years ago, but she got fed up with paying the cash to keep the patent going. Catherine – whatever you do, please don't read this next sentence . . . Over two hundred million fidget spinners have now been sold!

THE BIRO

In the 1930s, László Bíró invented a new type of pen with a tiny ball at the top and ink that didn't smudge, and he decided to name it after himself. That's right – he called it the Laszlo. Sorry, he called it the Biro. A company bought the patent off him for two million pounds, which sounds like a totally enormous amount of money . . . but this company then sold over one hundred billion biros! Think how rich László would have been if he hadn't sold up! ➤ Fact check – very rich. ⚡

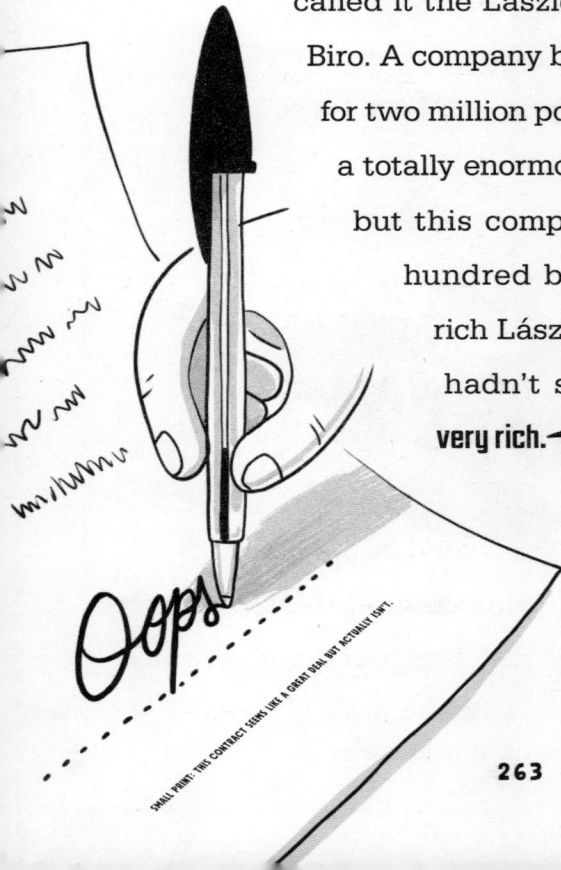

Oops

SMALL PRINT: THIS CONTRACT SEEMS LIKE A GREAT DEAL BUT ACTUALLY ISN'T.

ADAM KAY GENUIS ENTERPRISES LIMITED

ADAM'S CHAMPION CHOCOLATE DECKCHAIR

Don't you just hate it when you're lying down in the sun and have to get up for a snack? Well, that's a thing of the past when you buy the world's very first chocolate deckchair. Just gnaw on an armrest, then you can sunbathe and snack at the same time!*

Only £7,168.99 (white chocolate headrest £1,200 extra)
*Please note that the chocolate deckchair melts in warm weather, so can only be used in the winter.

TRAVEL: SPACE

With space, the clue's in the name. The universe is absolutely massive, and it's almost completely empty, dotted with a few planets and stars and asteroids. In fact, 99.9999999999999% of the universe is full of a big load of absolutely nothing. But the stuff that is there has fascinated humans ever since they wandered out of their caves, looked up into the sky and went 'Oooh!' at the moon. Let's see how we got all the way from there to jetting off to Mars.

WATCH THIS SPACE

We live in a delicious-sounding galaxy called the Milky Way, which is one of hundreds of billions of galaxies in the universe. The Milky Way contains thousands of different solar systems, including our one, which contains eight planets, all spinning round a big fiery thing called the sun. Stop me if I'm getting too technical here. We all learn that stuff quite quickly when we go to school, but for thousands of years humans living on Earth (that's the name of the planet we live on, by the way) had no idea about any of it.

They knew that daytime happened when the sun was in the sky, and that night came when the sun headed off, plus they drew maps of the stars they could see, and worked out that the moon changed shape throughout the month, but that was about it. One thing they were pretty sure about was that the Earth was at the centre of everything, and the other things they could see just spun round it. But I guess they couldn't look it up on Google yet.

This all changed when Nicolaus Copernicus came along. He published a book in 1543 called *Me and My Copper Knickers.* ➤ **Fact check - his book was called *On the Revolutions of the Heavenly Spheres.*** ➤ In it, he suggested that Earth actually goes round the sun, not the other way round. He'd actually thought of this theory about thirty years earlier, but was worried that his idea was so shocking he might get arrested or even killed for printing it. Copernicus died on the day his book came out – not from a spear through the spine though; he just collapsed.

HANS IN THE AIR

There's only so much you can learn about the sky just by looking with your eyes – you can check if it's a full moon, you can see a few stars, and you can tell if it's fireworks night. But if you want to know a bit more than that, you're going to need a telescope. The telesbloke (bloke who invented the telescope) was a Dutch optician called Hans Lipperhey. One day, in 1608, Hans saw a couple of children playing with a smashed-up pair of glasses, holding up two lenses, one in front of the other, so they could look at something far away. And what did Hans do? That's right – he had them sent to prison for ninety years. **Fact check - he used their idea to build the first-ever telescope.** The first telescopes weren't great at looking into space, but were pretty handy in battles for seeing where the enemies were hiding or for checking if the shop down the road had any Twixes left.

RUN AWAY! WE'RE BEING ATTACKED BY A GIANT WINGED BEAST!

MAGNIFICO GALILEO

Galileo Galilei was a very clever inventor. In fact, he was so good they named him twice. He was fascinated by what was going on up in space. He'd heard of Hans's telescope and thought it sounded amazing . . . but also slightly rubbish. GG reckoned he could do much better – why only gaze three streets away when you could be gawking at the stars? So he went off and worked on his own version, which could see five times as far as Hans's shambolic effort.

Galzo made loads of interesting discoveries with his exciting new toy. Everyone used to think that the surface of the moon was perfectly flat, like an extremely smooth bum – but he could see it was covered in craters and mountains, like a very spotty bum. He noticed that Jupiter had four little moons orbiting it called Weaselface, Bunion, Sploig and Bumfo. **Fact check – Jupiter's moons are called Io, Ganymede, Europa and Callisto.** He didn't spot the other ninety-one smaller moons that Jupiter has, but I'll let him off for that – it *was* four hundred years ago, after all.

People used to think that the Milky Way was just a big milky splat in the sky, but he had a closer look and discovered that it was made up of loads of tiny stars. And finally . . . he realized that every day he looked at

Venus it was a slightly different shape – going from a fingernail sliver all the way to a full circle, a bit like the moon. And the only way that this could happen was if the planets were rotating round the sun – he'd proved once and for all that Copernicus was right. Woohoo!

Well, not quite woohoo. The Pope got really annoyed, because the Church had always said that the Earth was at the centre of the universe, and no one likes to be wrong, do they? Well, apart from me – I don't mind being corrected. ➤ **Fact check - you got so cross about one of my fact checks that you threw your printer out of the window.** ➤ Poor G-Gal was put on trial, and even though he had his nice pictures of Venus as evidence, the Church weren't having it and just said he was lying. They imprisoned him in his house for the rest of his life, and all his books were banned, including any he hadn't even written yet. ➤ **Fact check - a recent survey found that 70% of people want your books to be banned.** ➤

HUBBLE BUBBLE

Locking up Galileo didn't stop people from wanting to know what was going on in the sky. Caroline Herschel was born in Germany in 1750 and worked as a singer. I wonder if my Great Aunt Prunella went to any of her concerts . . . Unfortunately, Caroline was an absolutely terrible singer, so her job didn't last very long. She then had a dramatic career change and became an astronomer.

It turned out that she was much better at using a telescope. Together with her brother William, she discovered the least funny planet, Uranus, and then she found eight comets and fourteen nebulae (clouds of space dust). The King of England, Adam IX ➤ **Fact check – George III. No kings have ever been called Adam.** was so impressed with Caroline that he gave her a job as an assistant astronomer, making her the first-ever female scientist to be officially paid for her work.

URANUS JOKES BANNED BY MY LAWYER, NIGEL

URANUS IS GOING TO MAKE US FAMOUS.

THANKS VERY MUCH!

WHY IS THERE TOXIC GAS COMING OFF URANUS?

BECAUSE I HAD BEANS FOR LUNCH.

SHALL WE SHOW URANUS TO THE KING?

OK, BUT I'D BETTER GIVE IT A WASH FIRST.

In 1946, a scientist called Lyman Spitzer realized that the layer of atmosphere between Earth and space was making any photo of space slightly blurry, like when

Pippin stole my phone and somehow managed to take a photo of my fridge. Lyman had a very simple idea for how to get better-quality photos – just put the telescope up in space. ➤ **Fact check - this was not a simple idea. The Hubble Space Telescope took forty-four years to get into space.** ➤ Next time someone complains that you're late with your homework, maybe remind them that the Hubble telescope took NASA forty-four years. (My lawyer, Nigel, has asked me to point out that this is a terrible excuse and definitely won't work.) The Hubby Telly looks like a normal telescope, but it's the length of a bus, with a mirror in the middle the size of a king-size bed. It was eventually flown into space in 1990 on a space shuttle called *Discovery*, but . . . it didn't work: all its images looked blurry and rubbish. Oops. It turns out that a mirror hadn't been polished correctly – by the thickness of a bum hair – and this was enough to ruin all the photos it took. Luckily a few years later they were able to send up some special specs to fit to the mirror and make it work properly. Phew!

A SIDE OF ROCKET

Looking at space was never going to be enough – what humans really wanted was to actually get up there. Some people think that the first person to have a go was a man called Wan Hu who lived in China about five hundred years ago. He was pretty keen to see what was going on above the clouds, but there was a big problem – spaceships hadn't been invented yet and he didn't have time to wait. So he got a bunch of massive fireworks, tied forty-seven of them to a chair and . . . BOOOOM! Up he went, and he landed on the moon just after lunchtime. ⚡ **Fact check - he exploded.** ⚡

No one really had much luck sending things to space until the 1950s, when the Space Race began. The Space Race was a bit like an egg-and-spoon

race, except there were only two participants –
the USA and the USSR (the old name for Russia
and some nearby countries) – and instead of
carrying an egg, each country tried to fly
things into space, and instead of being given
a trophy at the end they just got to say, 'We
won!' **Fact check – it was therefore nothing like
an egg-and-spoon race.**

The way we send things into space today
actually isn't very different from when Wan
Hu went to Fireworks Я Us. What a rocket
does is make an explosion, then push all
the fumes out in one direction. If you
imagine doing a fart so powerful that it
lifts you into the air, that's pretty much
it. Oh, by the way, it only counts as
space when you're sixty-two miles
above sea level, so you're unlikely to
get up there by farting alone.

Russia got off to a good start in 1957 by
sending the first two spacecraft into

space, Sputnik 1 and Sputnik 2. Sputnik 2 contained the first-ever passenger in space, a dogstronaut called Laika. Hang on, I'm just going to send Pippin out of the room. Laika was a stray dog who lived in Moscow and she became the first animal to ever orbit Earth, but sadly died a few hours into the mission. I don't Laika this story very much. OK, Pippin, you can come back in now.

The Americans weren't very happy to be losing the race, and were absolutely determined to be the first people to send a person into space. Then . . . oops. Russia won that as well, sending a man called Yuri Gagarin up there in 1961. Yuri orbited Earth for about an hour and three quarters, because that was how long it took him to watch *Toy Story 3*. ➤ **Fact check -** *Toy Story 3* **was not released until 2010.** ➤ An American astronaut called Alan Shepard got up into space three weeks later, but however much he said, 'First the worst, second the best, third the one with the hairy chest,' President John F. Kennedy wasn't impressed. JFK told NASA that they NTGSOTM (needed to get somebody onto the moon) in the next ten years. No pressure, then.

ABSOLUTELY OVER THE MOON

Going to the moon was a lot more complicated than NASA just putting 'the moon' into their satnav and driving off. It involved designing thousands of different things, like the world's biggest-ever rocket, a way to record their astronauts on the moon and transmit the images back to Earth, a heat shield so their spaceship didn't burn to dust the second they flew back into Earth's atmosphere, massive parachutes to get them safely onto dry land **⤙ Fact check - wet sea, not dry land. The lunar module splashed down in the Pacific Ocean.⤚** and building and programming the world's most complicated computer.

And it was all done by one person, who must have been absolutely exhausted by the end of it. **⤙ Fact check - four hundred thousand people worked on the moon mission.⤚** Yeah, that makes more sense, actually. One of the most important people involved in the Apollo 11 mission was Margaret Hamilton, who designed all the software that got the astronauts up to the moon.

It's time to use my robot butler's lie detector, so you can determine which of these facts about Margaret Hamilton is a total Brunei.

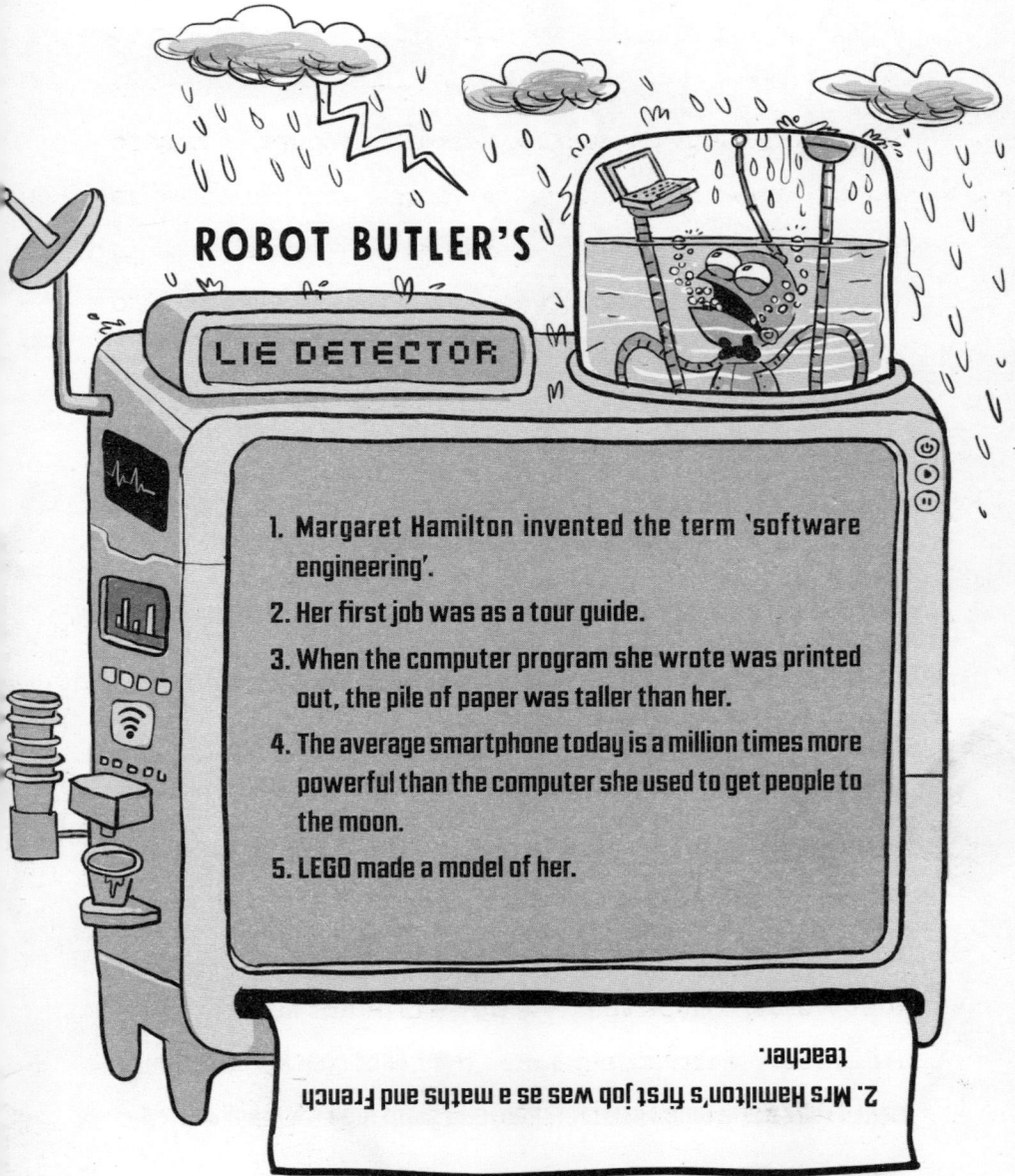

ROBOT BUTLER'S

LIE DETECTOR

1. Margaret Hamilton invented the term 'software engineering'.
2. Her first job was as a tour guide.
3. When the computer program she wrote was printed out, the pile of paper was taller than her.
4. The average smartphone today is a million times more powerful than the computer she used to get people to the moon.
5. LEGO made a model of her.

2. Mrs Hamilton's first job was as a maths and French teacher.

After years and years of research and practice missions, Apollo 11 launched from Cape Kennedy in America on July 16th 1969, and it landed on the moon four days later, which sounds like quite a long time, but I guess fifty thousand miles is a long way. Around 650 million people tuned in to watch Neil Armstrong become the first person to walk on the moon, saying his famous first words, 'I hope that one day a brilliant author called Adam Kay writes about me.' ⚡ **Fact check – he actually said, 'That's one small step for man, one giant leap for mankind.'** ⚡

Neil's colleague Buzz Aldrin joined him
on the moon shortly afterwards, and a third
astronaut, Michael Collins, had to stay behind
on the spaceship in case any Amazon parcels came.

The next plan is to send someone to Mars, and NASA
are hoping to fly a mission up there by the year 2035.
This might have already happened if you're reading
this book extremely slowly.

SUIT YOURSELF

If you're heading into space, you can't just wear your favourite tracksuit bottoms and Adam Kay sweatshirt – you need a spacesuit. Strangely, the space-suit was invented years and years before anyone actually went into space – a bit like if air freshener had been invented before the toilet. In 1935, Emilio Herrera wanted to take his hot-air balloon twelve miles up into the atmosphere, but he knew that previous attempts had failed because people had run out of something extremely important: biscuits. ➤ **Fact check - incorrect.** ✔ I meant Wi-Fi. ➤ **Fact check - try again.** ✔ OK, oxygen. So Emilio invented a suit made of rubber, wool and steel cables, all wrapped in silver – and when NASA were organizing their minibreak to the moon, they used Emilio's suit. Warning: please stop reading now if you don't like stories about wee going in people's eyes.

One thing Emilio and NASA hadn't thought about was what happens if someone really, *really* needs a wee. Well, Alan Shepard had been in his spacesuit for four hours when he desperately needed to go – and he was about two hundred thousand miles from the nearest public toilet. He had no option but to wee inside his spacesuit. It sloshed around and, because there wasn't any gravity, it went *everywhere*, so from then on astronauts have always worn big nappies when they're up in space. If you're wondering how toilets work on spaceships (of course you are) and why the poos don't just fly around the bathroom in zero gravity, it's because the spaceship toilet is basically a vacuum cleaner that sucks away everything that comes out of their astrobums.

There are lots of things we use down here on boring old Planet Earth that wouldn't be possible if people hadn't whizzed off into the atmosphere. If you want to know more, watch this *space*! Get it? ➤ **Fact check - my joke-assessment module informs me this has a 4% humour level.**

MEMORY FOAM

Memory foam mattresses remember the exact shape of your bum, so it's nice and comfortable every time you get into bed. But they were originally designed as double-comfy cushions for astronauts as they zoomed off to Jupiter.

⚡ Fact check – the moon, not Jupiter. Jupiter is made of gas, so it would be like trying to land on a cloud.⚡

HELLO AGAIN, STEVE.

INVISIBLE BRACES

NASA developed a see-through, super-strong material called translucent polycrystalline alumina to cover their antennae with. When dentists down on Earth were looking for a kind of dental brace that would be less obvious than the metal kind, they called their mates at NASA and borrowed a bit of translucent polywhatever-it-was-called.

GPS

GPS stands for 'Gorilla Puke Sandwich' **⚡ Fact check – GPS stands for 'Global Positioning System'.⚡** and it uses satellites to tell you where exactly you are on Earth, and the time, accurate to one billionth of a second. And it's thanks to all these satellites that we can use our satnav to get us from A to B, without getting lost and ending up at J . . . Satellites also take photos of the Earth to make maps, they tell us what the weather is like, and help us make phone calls. Look up to the sky and thank a satellite. Actually, maybe wait until you're on your own to do that – you might look a bit weird.

CAMERA PHONES

In the olden days, no one would believe you if you said you saw a hyena doing a handstand, or an explosion in a firework factory, or an extremely famous celebrity such as me. **Fact check – in the UK, one in 423,850 people has heard of you.** But these days you can prove the things you saw by taking a photo or a video on your phone – and that's all thanks to NASA, who developed this technology so that astronauts could take selfies on the moon.

SCRATCH-RESISTANT LENSES

If you're cool enough to wear glasses and clumsy enough to fall over the whole time, you'll be very glad that NASA invented scratch-resistant lenses. They originally made a scratch-proof coating so that astronauts would still be able to see out of their visors if bits of space grit flew into their helmets. But then opticians on Earth got jealous about this snazzy space glass, and – hey presto! – scratch-resistant lenses for everyone's glasses.

HANDHELD HOOVERS

Next time you're using a handheld Hoover, you can pretend you're on the moon collecting samples of rocks, dust and alien poo. Because that's what the first-ever Dustbuster was used for! ⚡ **Fact check - astronauts do not collect alien poo.** ⚡ Well, that's what they want you to think.

TRUE OR POO?

SOMEONE LANDS ON THE MOON EVERY EIGHTEEN MONTHS.

POO Well, not unless the Octopus People of Zaarg have been visiting there for their holidays. The last time anyone walked on the moon was over fifty years ago. Once the Space Race was over and the USA and Russia started working together, there wasn't much reason for anyone to go back. Can you blame them? The phone reception's awful up there.

A MONKEY, A GOLDFISH AND A SQUID HAVE ALL BEEN UP TO SPACE.

TRUE It's a bit annoying that a goldfish has been on a much more exciting holiday than I have. But so have a lot of vegetables – NASA is currently growing lettuces, cabbages and kale up on the International Space Station, to investigate how plants survive up there, and to make sure that astronauts eat their greens.

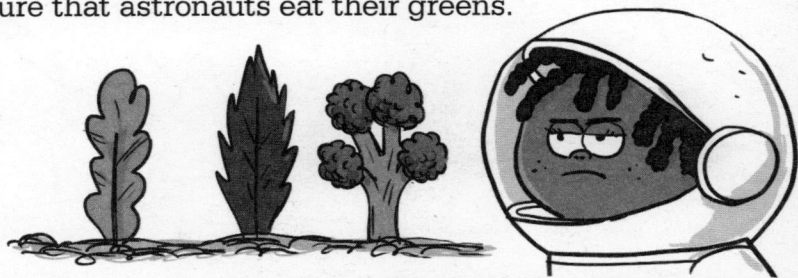

YOU'RE NOT ALLOWED TO BRING BREAD INTO SPACE.

TRUE No, the moon isn't allergic to gluten. It's because if you eat bread up in space, the crumbs float around everywhere and it's impossible to get rid of all of them. It's the same reason you're not allowed salt and pepper, or hundreds and thousands on your birthday cake. Apologies if this puts you off being an astronaut.

SO MUCH CHEESE AND NOTHING TO PUT IT ON.

ADAM'S ANSWERS

WHAT IS SPACE JUNK?

Well, Neil Armstrong and Buzz Aldrin absolutely loved Monster Munch, so the universe is totally littered with crinkled-up packets of pickled-onion flavour (Buzz's favourite) and Flamin' Hot (Neil's fave). ⤛ **Fact check – try again.** ⤜ OK, so the thing about space is that because it has no air, no atmosphere, no anything – whatever is left up there just kind of hangs around forever. Space junk means things like satellites that have stopped working, bits that have fallen off spaceships, and chunks of old rocket. Plus some lumps of frozen astronaut pee, poo and puke. There are currently over twenty thousand bits of junk hurtling around up there, so it's probably a good idea to wear a helmet next time you're in space. (You might need it for breathing too.)

WHAT DOES SPACE SOUND LIKE?

In 1932, a man called Karl Jansky couldn't understand why when he listened to the waves that his radio antennae were picking up, there was always a quiet hiss in the background. He checked that there weren't any deflating balloons or leaking bike tyres around – nope.

There weren't even hundreds of snakes under his bed. Then he realized that the sound was actually coming from space. I've had a listen to some recordings, and I'd describe space as sounding like one of those very long quiet farts.

WHAT DOES THE MOON SMELL OF?

Not cheese, so it can't be made of that after all. You can't smell the moon while you're standing on it because you'd be wearing a helmet, but an astronaut called Gene Cernan got back into his lunar module after a long day's moonwalking and had a sniff of his boots. And they smelled like . . . gunpowder. You might have smelled gunpowder before when fireworks have gone off, or even when you've pulled a Christmas cracker. I've got no idea why the moon smells of it though. Maybe Wan Hu made it up to the moon in his firework-chair after all?

FUNVENTIONS

Do you ever wonder where the toys in your house come from? No, I don't mean from your Uncle Clive. No, not from the toy shop. No, I didn't mean from the internet either. I meant who invented them? Well, have I got a treat for you! ⚡ **Fact check – only 2.3% of readers consider these sections to be a treat. Over 80% describe them as either 'boring' or 'annoying'.** ⚡

BALLS

Children have been kicking balls around for thousands of years, although they've changed a fair bit since then. The balls have changed, I mean, not the kids – they've always picked their noses and farted. In Ancient Greece, ball games were played with pigs' bladders. I hope they drained all the piggy wee out of them first, or heading the ball might have been a bit disgusting. Things got slightly better when Henry VIII (that's Henry the Eighth – his surname

wasn't pronounced 'Veeee') was around, five hundred years ago. Builders working in the Houses of Parliament found a ball he probably used to use, which was filled with mud and human hair.

JIGSAWS

The first jigsaw was made in 1762 by a mapmaker called John Spilsbury, to teach children about geography. He took a map, stuck it on some wood, cut it into bits and called it a 'dissected puzzle'. It was then renamed the jigsaw, after the kind of saw that's used to make them. Soon, grandparents all over the world were disappointing their young relatives at Christmas with yet another thousand-piece jigsaw when they actually wanted a Nintendo Switch.

URIN SCORE
6/10
SOUNDS SLIGHTLY DISGUSTING.

URIN SCORE
2/10
DON'T NAME A TOY AFTER A SAW.

LEGO

Ole Kirk Christiansen was a carpenter in Denmark who mostly made ironing boards and ladders, until one day in 1946 he had the idea of producing tiny little blocks of plastic that would fit together, and his lovely colourful bricks have been making barefoot grown-ups scream in agony ever since. Ole called his company LEGO, which comes from the Danish 'leg godt', which means 'play well'. There are now over 400 billion LEGO bricks on Earth, and about 390 billion of those are lying out on the carpet, even though you've been told that they have to be put away in their box.

THIS IS FOR LOSING MY ARM.

FIREWORKS

A couple of thousand years ago in China some people were trying to find the secret to eternal life by mixing all sorts of different chemicals together. They didn't discover the secret to immortality (sorry, spoiler alert), but they accidentally invented gunpowder. Gunpowder is so dangerous that it's basically the opposite of the secret to eternal life, but they stuffed it into tubes of bamboo and invented fireworks. Firework displays were just orange until the 1800s, when people discovered that you could add different metals to the gunpowder mix to make more exciting explosions! Want some gold sparks? Add iron. Fancy a loud bang? That'll be aluminium. A green colour? Get yourself some barium. Red? Head down to the strontium shop.

THESE WILL BE ADORED BY EVERY CREATURE ON THE PLANET!

SUPER SOAKER

People have been playing with water pistols for one hundred and fifty years – the first ones were made out of metal and you squeezed a rubber ball to fire the water. They were absolutely useless though – you'd get someone wetter by sneezing on them. (My lawyer, Nigel, has asked me to inform you that you should never sneeze on someone, as it is both rude and unhygienic.) That was until Lonnie Johnson came along. He was an African American engineer who used to work at NASA, sending satellites up to Jupiter. One day, he was doing an experiment when a nozzle accidentally blasted an extremely powerful jet of water across the room and he realized it would make an amazing water pistol. His invention worked by pumping the water so it was at a higher pressure, and it could squirt the length of a bus. Lonnie originally called it the Power Drencher but then he renamed it the Super Soaker.

URIN SCORE

5/10

NOT VERY CATCHY – SOUNDS LIKE A TYPE OF SHOWER ATTACHMENT.

URIN SCORE

8/10

EXCELLENT USE OF ALLITERATION. ALLITERATION, OR REPEATING THE FIRST LETTER OF WORDS, IS WHY THIS BOOK IS CALLED *KAY'S INCREDIBLE INVENTIONS* NOT *KAY'S QUITE GOOD DISCOVERIES.*

- Power Drencher ✗
- Mega Rinser ✗
- SOGGYTRON 5000 ✗
- SPLASHY SPLASHY FUNGUN ✗
- Wetminator ✗
- The Duke of Dampress ✗
- Sprinkle mageddon ✗
- Moist Maestro ✗
- Judy Drench ✗ Supersoaker ✓

PART THREE

TECHNOLOGY

MICE, MORSE CODE AND MARIO

Our ancestors who lived in caves tens of thousands of years ago simply wouldn't believe how we communicate with each other these days. For example, yesterday I wanted to see if my friend Bruce was free at the weekend, so I grabbed this tiny phone from my pocket and typed a quick message, then ten seconds later he replied to say that he wasn't because he was visiting his grandmother in Margate. **➤ Fact check – he was lying to you. ➤** Oh. Right. Well, let's see how we got from scrawling on walls to sending WhatsApps.

WRITE HERE, WRITE NOW

The very first kind of writing, back in the cave days, was more like drawing than actual sentences. So instead of saying, 'I'm going out to kill a cow, then cook it on a fire and watch *The Apprentice*,' they would draw a spear, a cow, a fire and Alan Sugar. **➤ Fact check – my image module suggests this book would be best represented as follows** 📖 🥱 🗑️. **➤** Historians argue about who the first people were to do 'proper' writing because historians love arguing about things. But most of them think it was the Sumerians, who lived in what is now Iraq, about five thousand years ago. Paper hadn't been invented yet, but that time period was known as the Bronze Age, so I'm sure you can guess what they used to write on. That's right, clay tablets. Their writing was very different from what we scribble today – it was loads of tiny little straight lines, written from top to bottom, rather than from side to side. When archaeologists found the oldest-ever Sumerian writing, everyone was extremely excited to find out what incredible things they had written so long ago. When language experts eventually decoded it all, it turned out to be a man complaining to a shop about the copper he'd bought from them.

It's very difficult to say when English first appeared, because language is changing all the time. For example, only this week, I've hugely improved the language with two brilliant new words: jarf (a jumper with an inbuilt scarf) and constipotato (when you've eaten so much potato that you can't poo). Around the year 450, Britain was invaded by a bunch of tribes from Germany and Denmark called the Angles, the Saxons and the Jutes – and a mishmash of all their languages ended up being known as Old English. Angle is where the word 'English' comes from, Saxon is where the word 'saxophone' comes from, and Jute is where the word 'juice' comes from. ➤ **Fact check – the saxophone is named after its inventor, Adolphe Sax, and 'juice' comes from the Latin word 'jus', meaning 'broth'.** ➤

Old English wasn't that much like the English we speak today. In fact, here's the first line from a famous poem from back then called *Beowulf*: 'Hwæt. We Gardena in geardagum, þeodcyninga, þrym gefrunon, hu ða æþelingas ellen fremedon.' Something about a garden, maybe? ➤ **Fact check – it translates as: 'So. The Spear-Danes in days gone by and the kings who ruled them had courage and greatness.'** ➤

HOW PAPER UNFOLDED

About two thousand years ago, a man called Cai Lun was working as an advisor to the Emperor of China. His job involved a lot of writing, which he had to do on bits of bamboo. Cai Lun was fed up – the bamboo was really difficult to write on, and very heavy, and they needed massive cupboards to store it in. So . . . he decided to invent paper.

AN ABSOLUTELY BRILLIANT
DRAWING OF SOME PAPER

£20 Million

Henry Paker

OH, THAT
HENRY PAKER
IS VERY
GOOD.

He chucked some wood, bark, old bits of cloth and fishing nets in a big pan of water, then mashed them to mush and spread it all out flat to dry in the sun. Then a couple of days later he had big bits of paper that he could cut up and write on. It seems like quite a faff. When I want paper, I just go to Ryman's.

PRESS HERE

I've been to a printing press to see one of my books getting produced and it was absolutely incredible: huge rolls of paper, bigger than oil barrels, are swallowed up by machines the size of buildings, which print, fold, cut and glue them, and books zoom out the other side. The company I visited could print over one million books in a day!

But rewind a couple of thousand years and every single copy of a book would have to be separately copied out by hand. This was usually done by skunks. Loads of skunks would sit in a room called a 'scriptorium' and – **Fact check - you mean monks.** Oh, yeah, that makes a lot more sense. Loads of monks would sit in a room called a 'scriptorium' and carefully write each book, word by word. They would also add illuminations, which were big fancy letters at the start of some sentences – maybe I should try some of those in my books.

ow does this look? Hmm, not great. There were a couple of major problems with copying books out by hand. First of all, it took absolutely ages to do – if you wanted to copy this book out yourself, it would take more than four days solid. And that's without any time to have a sleep or a sandwich or a wee. It was also extremely expensive to do this, and this meant that only very rich people could afford to own books, which was terrible because it stopped lots of people learning how to read.

This all changed when the printing press was invented. I don't know if you've ever done potato printing, where you carefully cut a picture of a bum into a potato (it doesn't have to be a bum, but bums are funniest), then dip it in some ink and press it onto some paper . . .

Well, the printing press was exactly the same idea, except, because potatoes aren't as big as books, they had to use blocks of wood. This was first done in China in the year 870, and then improved in 1300 by Wang Zhen, with a system that meant you could reuse the blocks afterwards by moving the letters into different places. He used his new invention to print a book he'd written about a young orphan who finds out that he's actually a wizard, then goes to school at a place where he meets two friends called Ron and Hermione and –

➤ Fact check - that's *Harry Potter*, which wasn't written for another seven hundred years. Wang Zhen's book was about new farming techniques.⚡

THIS BOOK IS RUBBISH.

Back then, there obviously weren't things like planes and email, so people couldn't find out what was going on in faraway parts of the world. This meant that no one in Europe knew about the printing press in China. In 1440, more than one hundred years later, a bloke in Germany called Johannes Gutenberg thought that he was being extremely clever by inventing a printing press. Johannes built it out of a contraption for squishing grapes to make wine, so I hope he cleaned it first. His version used metal letters rather than wooden ones, but was basically the same as Wang Zhen's idea. It's a bit like when I invented this amazing brand-new way of using bread and jam for a delicious snack, but it turned out that someone else had already invented the sandwich.

The most famous book that Johannes printed was the Gutenberg Bible. There are still a few copies around today – why don't you check your bookshelf now to see if you've got one? I'll wait.

Any luck? No? That's a shame, because they're worth about twenty million pounds each. You'd definitely know if you had one, though, because they're massive and very heavy – weighing about as much as a husky or a microwave or seventy cans of Coke. In fact, a man once tried to steal a copy from Harvard University in America and ended up falling and breaking his leg and his skull because it was so huge.

NOT MY TYPE

Not all inventors make something because they think it's going to change the world or earn them a Brazilian pounds. **➤ Fact check - 'Brazilian' means someone from Brazil. It's not a number. ➤** Often they're just trying to make something useful for them or their friends. And that's what happened in 1802 with Pellegrino Turri and his friend with a tongue-twister name, Countess Carolina Fantoni da Fivizzano. Countess Tongue-Twister had recently become blind and was finding it very difficult to write letters, and Pellegrino wanted to help. So he went to his workshop and rustled up the world's first typewriter – a machine that stamps letters onto a piece of paper. This is why it's always great to be friends with an inventor, and why I'm so popular. ➤ Fact check – your friend Bruce made up an excuse so he didn't have to see you. ⚡

HENRYISAVERY
COOLGUYWHER
EASADAMKAYS
MELLSTERRIBLE

The first typewriter that you could buy in the shops was made in 1873 by a company called Remington, who also made guns – very handy for murderers who like writing poems. Some of the improvements they made are still used on computer keyboards today. For example, they added a shift key, which literally shifted the keyboard to the right and changed the letters from lower case to SHOUTING. And in the English language they put the letters in the order that keyboards have today, known as the QWERTY keyboard, because of its first six keys. If you're in France (*bonjour!*), it's the AZERTY keyboard. Or perhaps you're in Germany (*guten tag!*), where it's QWERTZ. Or maybe in Ukraine (привіт!), where it's ЙЦУКЕН. Or in Zaarg (🐙🐙) where it's 🐙🐙🐙🐙🐙🐙. An interesting fact is that 'typewriter' is the longest word you can type only using the top row of a QWERTY keyboard. ➤ **Fact check - incorrect. There is a type of plant called 'rupturewort'.** ➤

USE THE MORSE, LUKE

Before the phone was invented, it was difficult to send messages over long distances. You could use smoke signals, but what would happen if it's foggy? You could have a pigeon fly there with a message strapped to its leg, but what if it got lost or eaten by an eagle? You could play drumbeats, but what if there was a thunderstorm or a loud Beyoncé concert nearby?

Samuel Morse worked as a painter. Not the kind who paints your bedroom walls but the kind who draws boring portraits of old men with big moustaches. Then in 1837 he came up with a great idea for sending messages across long distances. I'm not sure why – perhaps he wanted to tell someone in another city that he'd done another boring portrait of an old man with a big moustache? Sam's plan was very simple: he gave every single letter of the alphabet a code made out of dots (short beeps) and dashes (long beeps). Someone would tap a message in this code using a button, which would push down and complete an electrical circuit, and send a signal down a long wire. Then, at the other end of the wire, those little pulses of electricity would make beeping sounds. Here's a diagram or, if that's not your thing, a picture of Wolverine eating baked beans.

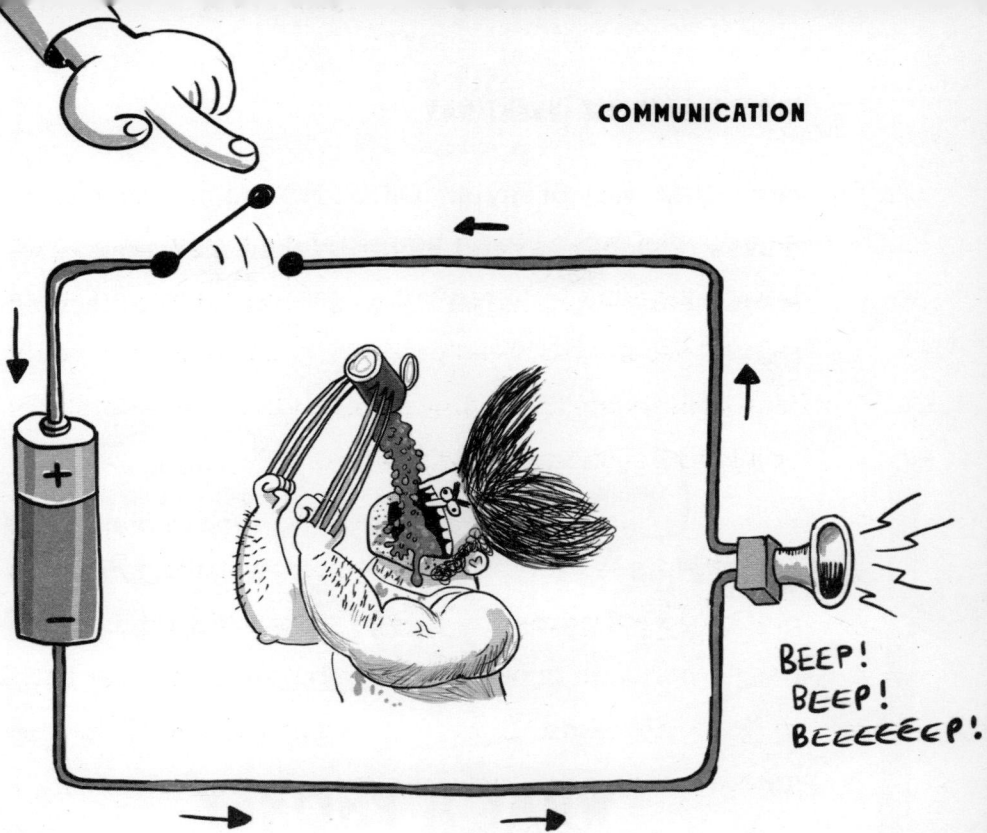

BEEP!
BEEP!
BEEEEEEP!

Letters that weren't used so much had longer codes; for example, F is $\cdot\,\cdot-\cdot$ and letters that were used more often had shorter codes, so A is $\cdot-$ and R is $\cdot-\cdot$ and T is $-$

Can you think of a word that uses those letters? That's right, RAFT would be $\cdot-\cdot\quad\cdot-\quad\cdot\,\cdot-\cdot\quad-$ What do you mean, you were thinking of a different word? Disgusting.

In 1844, Samsam installed a thirty-eight-mile wire in America from Washington to Baltimore, and sent the first long-distance message, or 'telegraph' as it was known. It went 'What hath God wrought', which was an

311

olden-days way of saying 'OMG'. For the first time ever, it was possible to send long-distance messages, and soon there were telegraph wires zigzagging all across the · — — — · — · · — · · — · ·

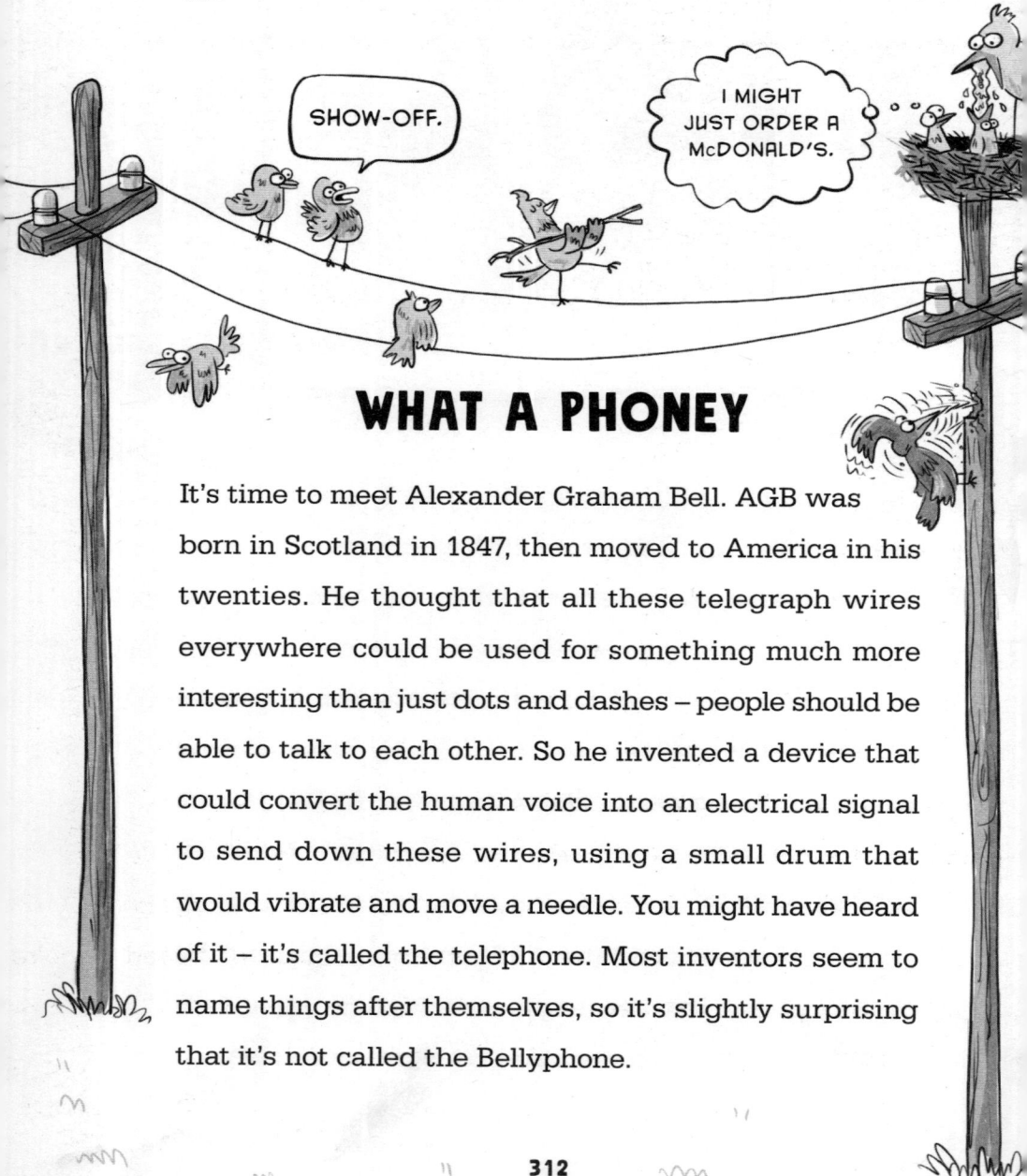

SHOW-OFF.

I MIGHT JUST ORDER A McDONALD'S.

WHAT A PHONEY

It's time to meet Alexander Graham Bell. AGB was born in Scotland in 1847, then moved to America in his twenties. He thought that all these telegraph wires everywhere could be used for something much more interesting than just dots and dashes – people should be able to talk to each other. So he invented a device that could convert the human voice into an electrical signal to send down these wires, using a small drum that would vibrate and move a needle. You might have heard of it – it's called the telephone. Most inventors seem to name things after themselves, so it's slightly surprising that it's not called the Bellyphone.

AGB made the world's first-ever phone call on March 10th 1876. Who do you think he called to demonstrate this world-changing discovery? The President? The Queen? The Pope? Nope, none of those. He was in his laboratory and accidentally spilled some acid on his leg, so he phoned his assistant in the room next door to come and help him. He probably said something like, 'Aaaaaagh! Aaaaaagh! Help! There's acid on my leg!' (My lawyer, Nigel, has asked me to remind you not to spill any acid on your leg.)

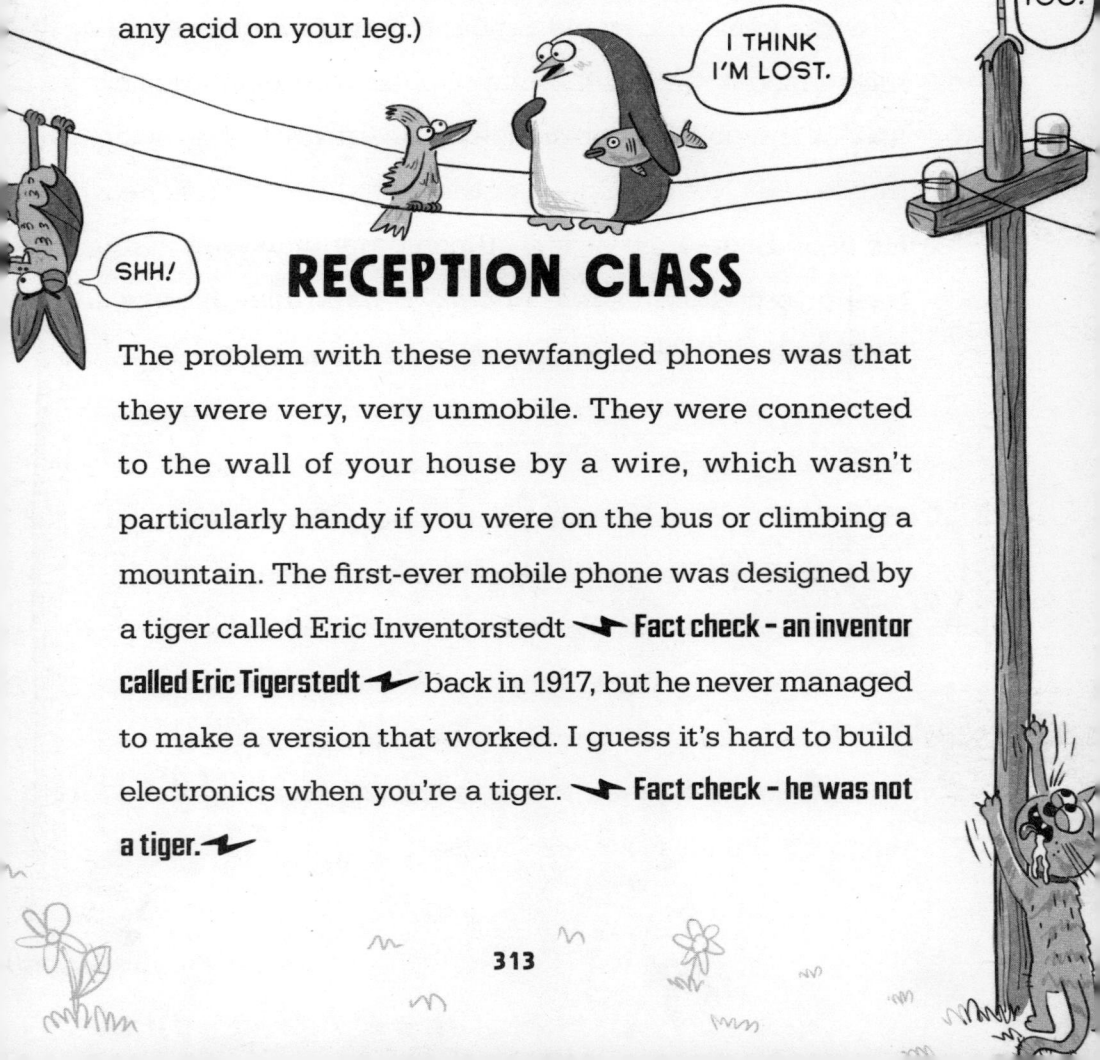

ME TOO.

I THINK I'M LOST.

SHH!

RECEPTION CLASS

The problem with these newfangled phones was that they were very, very unmobile. They were connected to the wall of your house by a wire, which wasn't particularly handy if you were on the bus or climbing a mountain. The first-ever mobile phone was designed by a tiger called Eric Inventorstedt ➤ Fact check - an inventor called Eric Tigerstedt ➤ back in 1917, but he never managed to make a version that worked. I guess it's hard to build electronics when you're a tiger. ➤ Fact check - he was not a tiger. ➤

The first person to make a working mobile that people could buy in the shops was called Martin Cooper. He worked for a company called Motorola and in 1973 he invented the DynaTAC 8000x, which sounds like where Darth Vader lives, but was actually a massive phone. It was the size of a shoe and weighed six times more than an iPhone, and the battery only lasted thirty minutes – although it was so heavy that if you held it for thirty minutes your arm would probably fall off. People even nicknamed it 'the brick'. But it worked! Martin made the first-ever mobile-phone call on April 3rd 1973 – but who did he call? Not the Queen or the President or the Pope. He called someone at a rival company who was also trying to make a mobile phone, to basically just say, 'Hahahahaha!' I'd have probably done the same.

EVERYONE WILL BE SO JEALOUS OF MY NEW MOBILE PHONE.

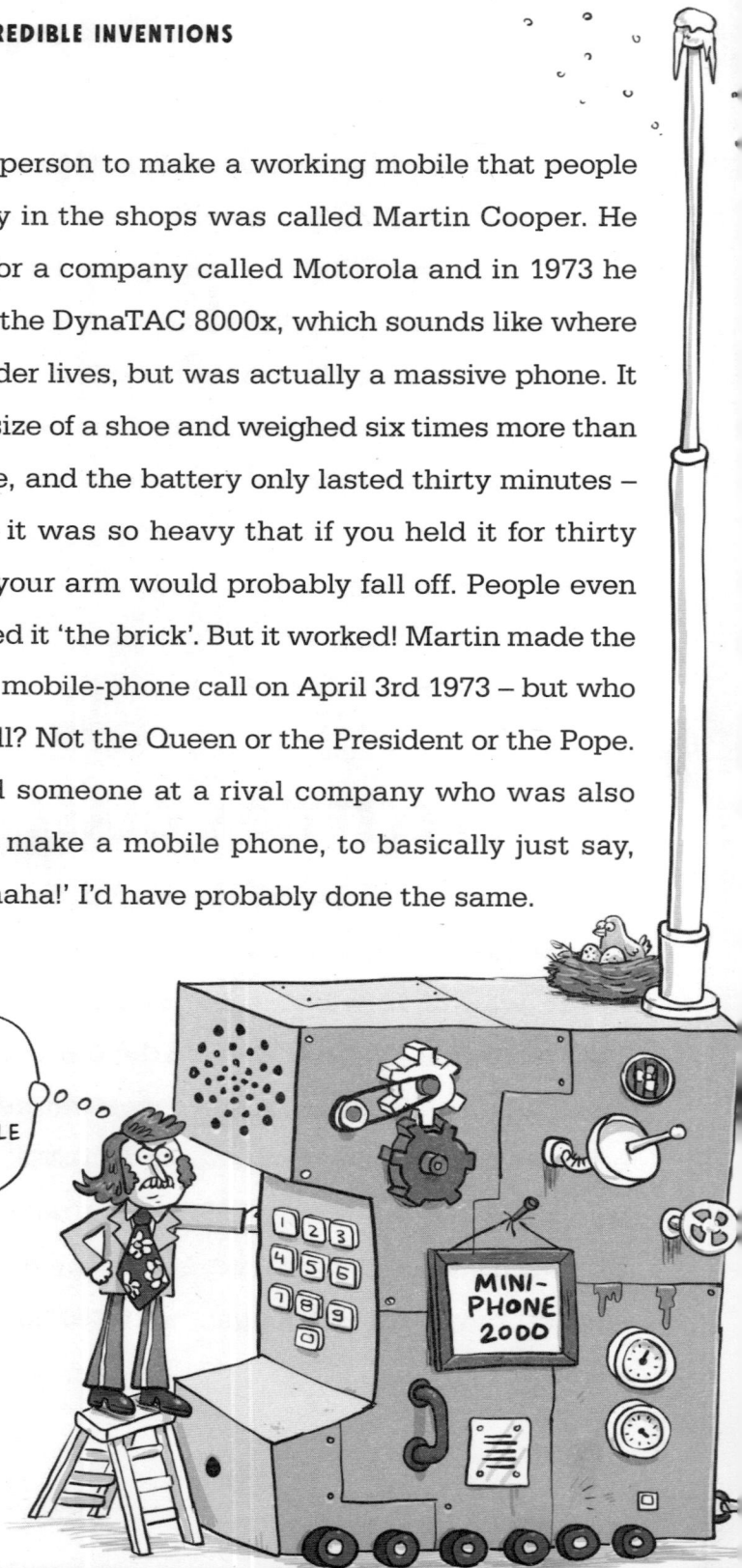

MINI-PHONE 2000

Mobile phones still work by turning noises into a signal, but instead of sending that signal down a wire, they blast it into the air using radio waves. Old mobile phones even had an aerial that you'd have to pull up to make them work. These radio waves get picked up by a big phone mast, which sends out much stronger radio waves to other masts, and these bounce around until they hit the mast near the person you're calling. And you don't hear any delay from all this bouncing around because radio waves travel at 671 million miles per hour (also known as the speed of light).

THE TEXT BEST THING

The first mobile phones could only make calls – they didn't have anything like a camera or a calendar or TikTok. You couldn't even send text messages for nearly twenty years! The first kind of texting was called SMS, which stood for 'Short Message Service'. SMS messages really were short, with a maximum of 160 characters. This is how long a 160-character message would be: 'Hi Bruce, I wonder if you might be free to hang out this weekend? It's been ages since I've seen you and I

thought we could maybe go to the cinema or visit the –' The first-ever SMS was sent on December 3rd 1992 by an engineer called Neil Papworth and it read: 'Merry Christmas'. Like I said, they didn't have calendars on their phones back then.

Because these phones didn't have touchscreens or even keyboards, you had to write your messages using the number buttons, which took absolutely ages. Each number on the phone had three or four letters written next to it – for example, the number 2 had ABC, 6 had MNO and 8 had TUV. If you'd just had a nice avocado and you wanted to text the word 'AVO', you would type the numbers 2-8-6. Unfortunately, phones wouldn't always guess the right word, so it might write a different word you can also type with the numbers 2-8-6, such as BUM. Wouldn't that be terrible? Emojis weren't a thing yet either, so you'd have to make them yourself using normal letters. So a smiley face would be :-) Sticking your tongue out would be :-p and a zombie would be »¬°-°«¬

In 2002, MMS (multimedia messaging service) came along, which meant that people could now send really long messages about my books, pictures of my books, videos about my books and emojis about my books.

➤ **Fact check – my image module suggests the following emoji:** 💩 . ➤ How rude. This led to services like WhatsApp, which now send over one hundred billion messages a day. That's about twelve messages a day for every single person on Earth!

SMARTY-PANTS

In 1996, Nokia introduced a phone called the 9000 Communicator. Anyway, it looked like a normal (but

317

slightly massive) mobile phone with buttons and a screen on the front, until you clicked a switch on the side and it would open up like a book. Inside it had a proper keyboard and a bigger screen, so you could send emails and browse the web, if you didn't mind a web page taking about fifteen years to load. The first colour screen came in 1998, the first camera phone arrived in 1999, and the first iPhone was released in 2007. Grown-ups now spend over three hours of every day on their smartphones, which is disgraceful. They should be more like me and spend that time going outside and reading books and –➤ **Fact check – yesterday you spent over nine hours playing *Candy Crush*.**◀ Shhh!

LIGHTS, CAMERA, PHOTOGRAPH!

It's never been easier to take a photo of someone – just press a couple of buttons on a phone and it's up on Instagram before they can even say, 'Who are you?! Stop taking pictures of me!' In 1824, it was a bit trickier. A French man called Nicéphore Niépce worked out a way to take the first photo, using a tiny pinhole and a piece of metal covered with petrol and lavender oil. The only problem was that it took at least half a day to take the picture, which is quite a long time to be smiling at a camera. Oh, and the photo quality was totally awful. ➤ **Fact check – that's two problems.** ➤ His friend Louis Daguerre improved the technique so it only took a minute or two to take the photos, although he called them Daguerrotypes. Before long, loads of people were having their portraits taken and seeing pictures of themselves for the first time ever. And then, if they were anything like me, they would moan that they looked awful in them and blame the photographer. For the next hundred years, if you took a photo of someone, it would be stored on a roll of film inside your camera, which you would then have to take to a shop to get them to print out into physical photos.

URIN SCORE
2/10
LONG, DIFFICULT
TO REMEMBER
AND BOASTFUL.

In 1947, a three-year-old called Jennifer Land asked her dad why the process was so long and boring – she wanted to see the photo he'd just taken *now*. Her dad, Edwin, agreed, and went off and invented the Polaroid camera, which shot out a little photo developed by clever chemicals inside it. Soon, digital cameras revolutionized photography all over again, letting people see their photos on a screen as soon as they were taken. It also meant that people could edit their own images for the first time. This can be useful, but it wasn't very funny when I discovered that someone had Photoshopped Pippin's bum where my mouth should be on all the Christmas cards I sent out last year. I just wish I could work out who it was. **Fact check – no comment.**

THE WORLD AND HIS WI-FI

You might not have heard of Hedy Lamarr, but in the 1940s she was one of the most famous people in the world, starring in all the biggest films. **Fact check – this book is called *Kay's Incredible Inventions*, not *Kay's Fantastic Film Stars.*** Yes, thank you, I'm just getting to that.

She found acting a bit boring, so she'd go home every day and invent things.

During World War II – that's two, not eleven – **Fact check – 99.3% of your readers know that.** the American army was having a problem shooting its missiles, because the German army was blocking the radio waves that guided them. Hedy came up with an invention called 'frequency hopping', which meant that the radio signals would change all the time, which made it impossible for them to get blocked. And this system is still used today in Wi-Fi, which also uses radio waves to send information. If it wasn't for Hedy, our computers would all still be plugged into the wall.

Let's give my robot butler's lie detector another spin, so you can work out which of these facts about Hedy Lamarr is a massive shepherd's pie.

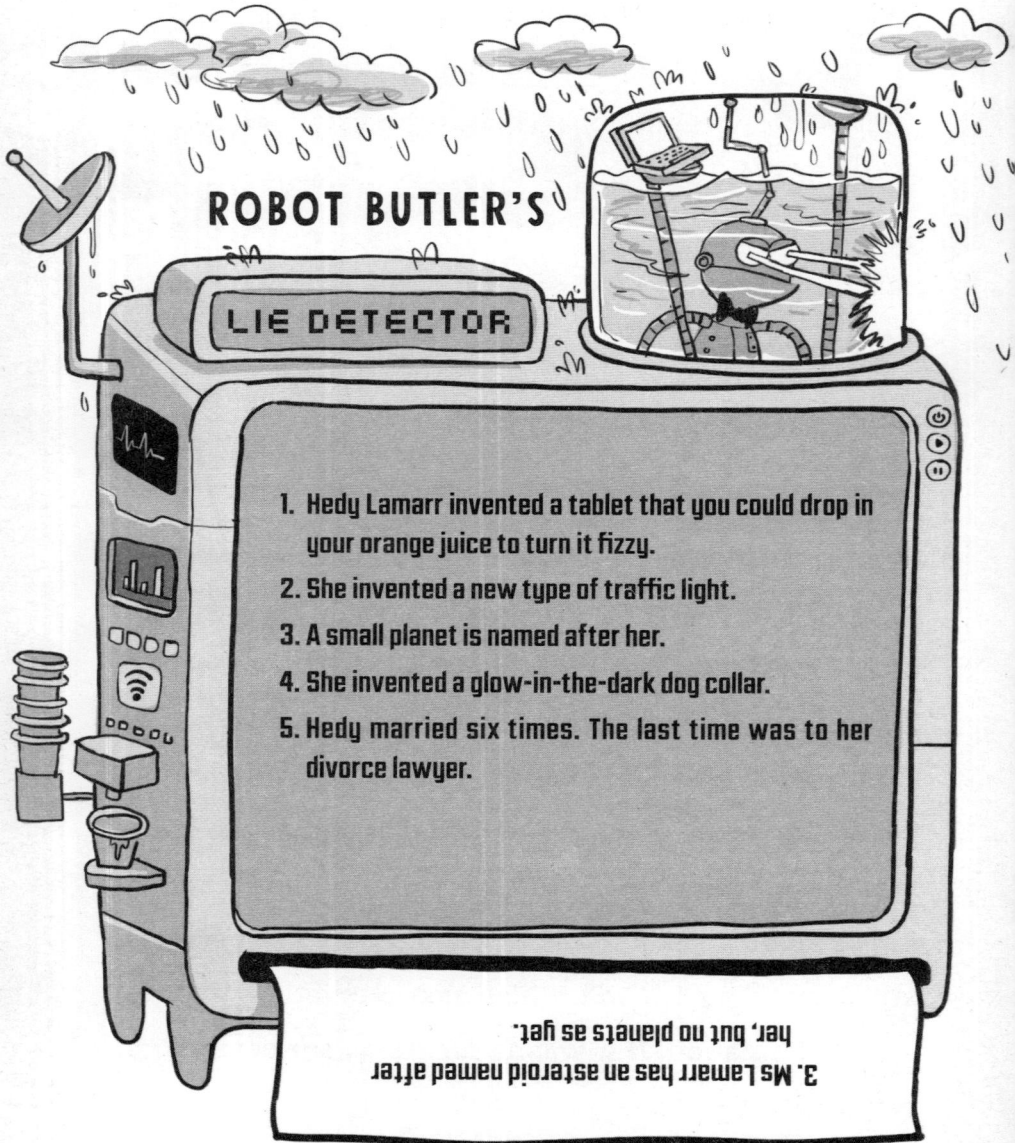

ROBOT BUTLER'S

LIE DETECTOR

1. Hedy Lamarr invented a tablet that you could drop in your orange juice to turn it fizzy.
2. She invented a new type of traffic light.
3. A small planet is named after her.
4. She invented a glow-in-the-dark dog collar.
5. Hedy married six times. The last time was to her divorce lawyer.

3. Ms Lamarr has an asteroid named after her, but no planets as yet.

TRUE OR POO?

WHEN THE TELEPHONE WAS FIRST INVENTED, PEOPLE WOULD ANSWER BY SAYING, 'AHOY-HOY!'

TRUE When the phone was invented, people didn't really know what to say when they answered, so Alexander Graham Bell suggested 'Ahoy-hoy!' People clearly thought it sounded a bit silly, so that didn't last long. Thomas Edison came up with a new thing to say . . . 'Hello!' Before Tommy Eddy, 'Hello' was previously something people only said when they were surprised.

THE POO EMOJI IS THE MOST COMMONLY USED.

POO   I mean that's false, not that the answer is the poo emoji. Sadly, it's only ninety-ninth on the list – second place is the crying-face emoji, and top spot goes to the crying-with-laughter emoji. My most-used emojis are the dog emoji and the puff-of-air emoji, when I need to warn people not to come into the living room because Pippin has farted.

I THINK HE'S SPENDING TOO MUCH TIME ON HIS PHONE.

THERE ARE MORE PEOPLE IN THE WORLD THAN MOBILE DEVICES.

POO   There are about sixteen billion smartphones and tablets in the world connected to the internet, and around eight billion people. If you're interested, there are also about one billion dogs, one and a half billion cars and, most importantly, three million people called Adam.

ADAM'S ANSWERS

HOW FAST IS IT POSSIBLE TO TYPE?

The world record is held by Barbara Blackburn, who could type 212 words per minute, which is five times faster than most people and means she could type this whole chapter in eighteen minutes! If you're interested in knowing the person who's fastest at typing on a keyboard with their nose **→ Fact check – this is the first thing in the book that over 90% of readers are interested in. ↙** the record goes to Davinder Singh, who nose-typed a seventeen-word sentence in forty seconds.

WHY HAS BLUETOOTH GOT SUCH A SILLY NAME?

**URIN SCORE
3/10
RIDICULOUS,
SOUNDS LIKE A
TYPE OF
DENTAL FLOSS.**

It's named after a Viking king from a thousand years ago who was known as Harald Bluetooth. The technology got this name because Harald persuaded lots of different tribes to talk to each other during his reign as king, a bit like how Bluetooth brings together my phone, my wireless printer and my fridge.

WHY DO PHONES ON APPLE ADVERTS ALL SHOW THE SAME TIME?

When Steve Jobs, who started Apple, introduced the iPhone for the very first time, it was 9:41 a.m. He thought that the time was lucky, so every advert for an iPhone, iPad or Apple computer ever since has shown 9:41 a.m. on the screen. More than two billion iPhones have been sold since, so maybe it worked?

Oh, I've just had a text from my friend Bruce. He's free to go to the cinema this weekend after all.

I *MUST* REPLY TO ADAM.

NOMENTIONVENTIONS

There's absolutely nothing worse (except for eating mushrooms) than doing loads of hard work and not getting any credit for it. Like how no one believes that I invented peanut butter. **↘ Fact check - peanut butter was invented by Marcellus Edson, one hundred years before you were born.↙** Throughout history, people have invented things that have changed the way we live, but have ended up getting forgotten. Often they were brilliant women who were shoved out of the way by greedy men who wanted everyone to think that they were the geniuses instead.

ELIZABETH MAGIE

I don't know if you've played Monopoly? It's a board game where you go round buying houses and charging rent. Yeah, it doesn't sound great when I describe it like that, but it's really popular and over 250 million sets have been sold. The person who took the credit for it, and all the money as well, was a bloke called Charles Darrow. But he'd actually stolen the idea from a game that already existed, made thirty years earlier, in 1904, by a woman called Elizabeth Magie, who only ever earned a few hundred pounds from it. How mean! But thanks to Elizabeth, people all over the world can have enormous arguments, flip over the table in fury and stomp off to their bedroom screaming, 'It's not fair!'

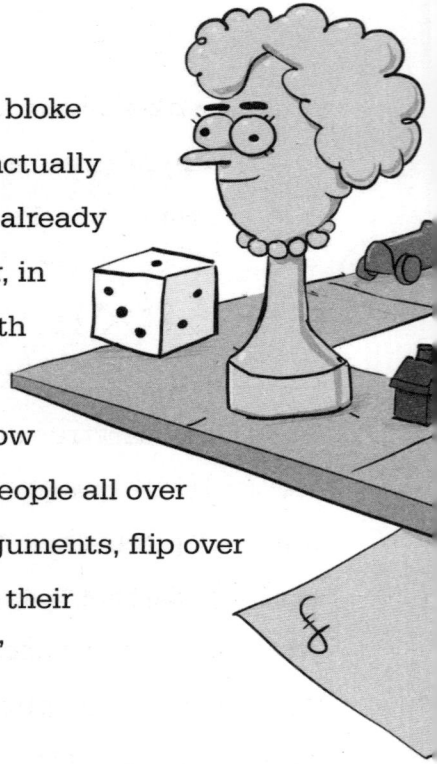

ROSALIND FRANKLIN

When I told you about DNA, you might remember I said that the shape was discovered by four people called Rosalind Franklin, James Watson, Francis Crick and Maurice Wilkins. I put Rosalind Franklin first on that list because she was the first person to produce pictures of DNA, which was the most important step in working out its structure. But she didn't get any credit for it at the time. In fact, the three blokes all won Nobel Prizes for their work, and she didn't! Ridiculous! Luckily these days everyone knows about her important

work. She's got lots of things named after her, including seven universities and laboratories, and even an asteroid called 9241Rosfranklin.

EUNICE FOOTE

The greenhouse effect is the lovely difference you can make to your street by painting your house green.

➤ Fact check - the greenhouse effect means that certain gases in the atmosphere, such as carbon dioxide, result in the Earth heating up, and is why burning oil and gas leads to global warming. ➤

The first person to spot this was Eunice Foote, in 1856. But – surprise, surprise – no one paid any attention to her, and someone else got all the credit. Another scientist called John Tyndall became famous for 'discovering' the greenhouse effect three years later, and it wasn't until ten years ago that people realized that Eunice had got there first.

MARY ANDERSON

In 1903, cars were pretty new on the scene and they still had quite a few issues. One big problem was what happened if it rained or snowed – drivers would have to pull over at the side of the road all the time, get out a cloth and wipe the windscreen.

A woman called Mary Anderson thought she could do a little bit better than that, so she invented windscreen wipers. For some reason, car companies weren't interested at all – maybe they thought people liked getting out of their cars every thirty seconds? But in 1967, a man called Robert Kearns came up with the same design with one little change – the wipers would come on every few seconds rather than swooshing away constantly. His version became really popular, and now everybody thinks that he was the person who invented windscreen wipers! Well, everybody except for you.

NEW FROM

ADAM KAY GENUIS ENTERPRISES LIMITED

ADAM'S SPECTACULAR SELF-WASHING TROUSERS

Doing the laundry is such a hassle, isn't it? Well, thank goodness for the world's first-ever trousers that wash themselves! Using a system of hoses and a handy soap dispenser, this ingenious wardrobe essential will never need to go in the washing machine.*

Only £897.99 (available in fluorescent green or diarrhoea brown)
*Please note that when your trousers are self-washing, it does look quite a lot like you've weed yourself.

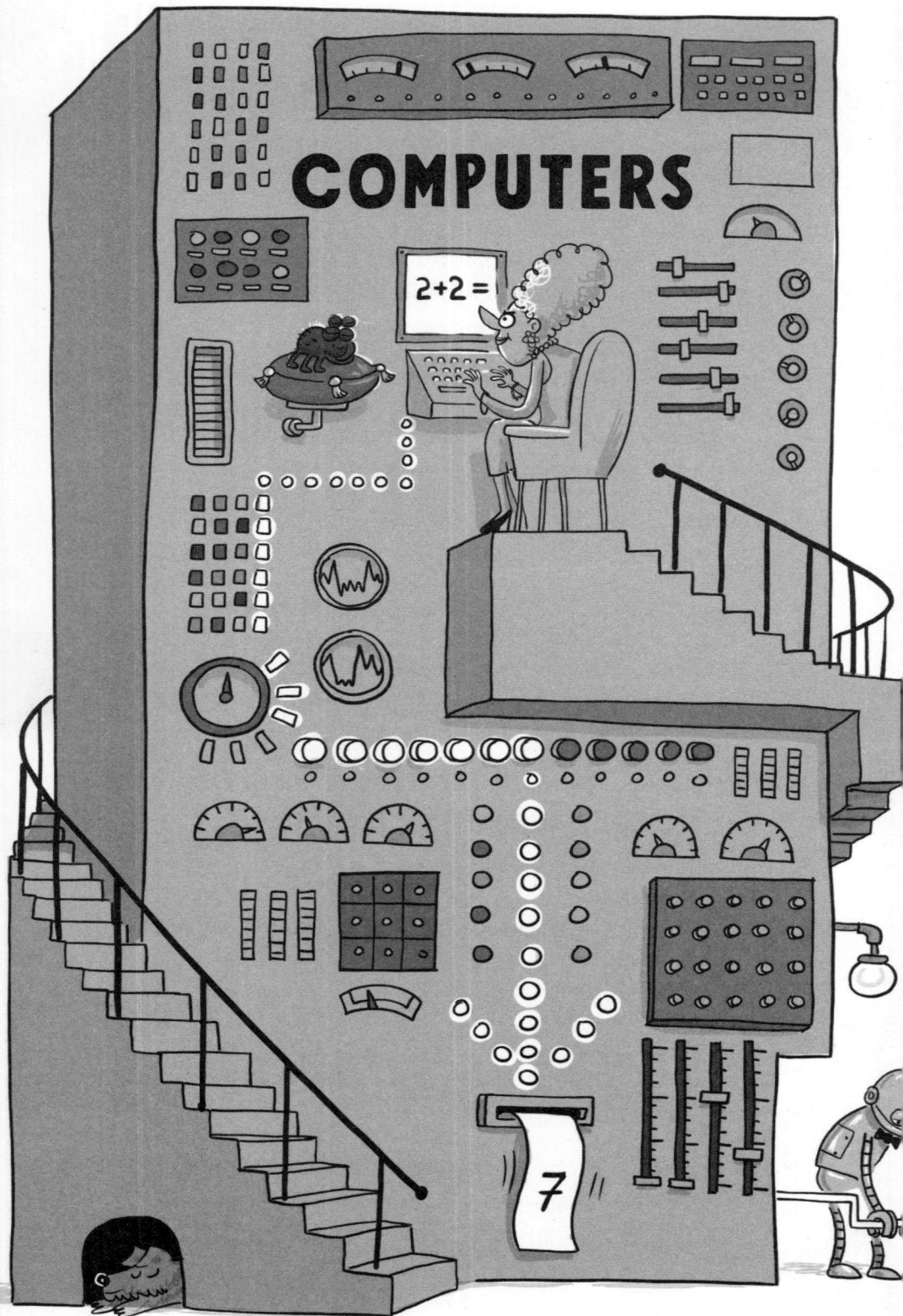

The computer is a member of the Cucurbitaceae family. Originally from South Asia, it grows on the ground as a vine and is known for its long, cylindrical green fruit and – ➤ **Fact check - that is a cucumber.** ➤ Oh yeah.

Computers are in every single aspect of our lives now. From the cars that take you to school to the traffic lights that get you safely there, to the cash machines we get money from, to the shops we spend it in, to playing computer games, to the laptop I wrote this book on. Although that has been slightly malfunctioning ever since Pippin weed on it. Let's find out about all the brainboxes behind computer boxes.

YOU ABSOLUTE LOOM

Here's a quick quiz. How long have computers been around?

A) 2,000 years

B) 200 years

C) 20 minutes

D) I don't know – I picked this book up by mistake . . .

The answer is B – if you got that right, then you win a free aeroplane. Please collect it from your nearest airport. (My lawyer, Nigel, has asked me to point out that your nearest airport is very unlikely to give you a free aeroplane, so it's probably not worth going.)

The first computer was built in 1801 by a French weaver called Joseph Marie Jacquard. It wasn't the sort of computer you'd recognize today. It didn't have a screen or a keyboard – so you couldn't even play games like *Attack of the Zombie Potatoes* (new from Adam Kay Genuis Enterprises Limited, only £487.99).

Joseph's computer was actually a type of loom, which is a machine that weaves fabric. He had a factory that made patterned silk cloth, but it was a real faff to make each design: a red bit, then a green bit, then a jangle-bangle bit, then a white bit . . . So he designed wooden cards, with holes punched in, that the loomputer could read and then make the right designs. He'd basically written the first-ever bit of code – and that means it was the first computer. Whatever-the-French-for-congratulations-is, Jojo! **➤ Fact check – the French for 'congratulations' is 'félicitations'.◄**

EXCESS BABBAGE

Charles Babbage was an extremely clever man, and there wasn't much he couldn't do, except for stopping people from calling him Mr Cabbage, which must have been pretty annoying. He was a mathematician, an inventor, a philosopher and a politician, but he was best known for building one of the world's first computers. Sorry, my laptop's been doing stuff like that ever since the wee incident.

It was the 1800s and Charlie was fed up with doing all his sums by hand. The easy ones weren't a problem, because everyone knows that $4 + 3 = 9$ ⚡ **Fact check - ahem.** ⚡ but CharlieBabs was doing really complicated sums, like ones to work out which direction a ship needed to steer to cross an ocean, or what size beams to put in a house so it didn't fall down. Not only did these calculations take absolutely ages, but it's very easy to make mistakes when you do things by hand, if you're tired or hungry or your robot butler interrupts you by accidentally spilling a load of soup on the floor. ⚡ **Fact check - that was not an accident.** ⚡

So ChazBabz invented a computer called the Difference Engine. I'd be exaggerating if I said it was portable. It was bigger than your bedroom (unless you live in a palace, in which case can I have a crown, please?) and it weighed as much as two cars. If you wanted to do a sum, then you'd enter the question by spinning some big wheels, then you'd have to turn a handle, and loads of cogs and chains and gears and axles would spin and whizz and whirl, and it would tell you the answer. I don't know why he didn't just use the calculator app on his iPhone, to be honest. ⚡ **Fact check - it was 1821.** ⚡

It's time for another go on my robot butler's lie detector to see if you can detect which of these facts about Charles Babbage is a complete bonsai.

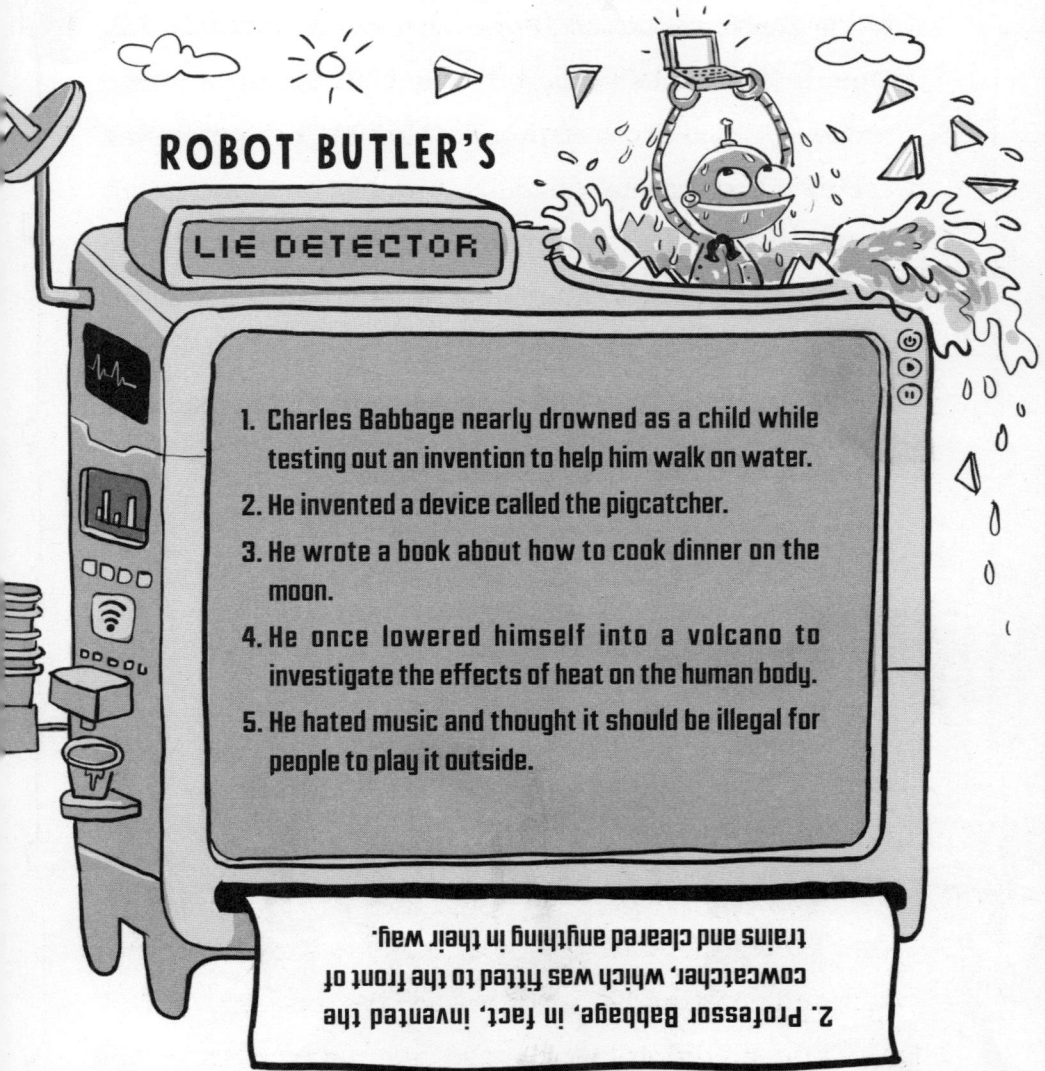

ROBOT BUTLER'S

LIE DETECTOR

1. Charles Babbage nearly drowned as a child while testing out an invention to help him walk on water.

2. He invented a device called the pigcatcher.

3. He wrote a book about how to cook dinner on the moon.

4. He once lowered himself into a volcano to investigate the effects of heat on the human body.

5. He hated music and thought it should be illegal for people to play it outside.

2. Professor Babbage, in fact, invented the cowcatcher, which was fitted to the front of trains and cleared anything in their way.

LIVE, LAUGH, LOVELACE

If it wasn't for Ada Lovelace, there would be no *Revenge of the Zombie Potatoes*. (Pre-order now from Adam Kay Genuis Enterprises Limited, only £882.99.) Ada was a friend of ChazBabz and thought that his computer was pretty nifty, but she reckoned it would be an awful lot better if someone wrote some programs for it. She remembered about Joseph Marie Jacquard's loomputer with its punched wooden cards and thought there was no reason she couldn't do the same sort of thing for this new invention.

Unfortunately, there was one big reason. A bunch of horrible men thought that women weren't as clever as them – which is obviously 100% wrong, and is known as sexism. This meant that they weren't interested in Ada's ideas and wouldn't even let her use the library she needed for her research. Ada just shrugged her clever shoulders and got on with her work at home. And did she manage it? Of course she did. She wrote a program that worked for CBabz's computer, and invented coding. She even predicted that one day computers would be able to write music and solve puzzles, which was pretty impressive guessing. Although she didn't predict *Revenge of the Zombie Potatoes*, which would have been more impressive.

BEFORE, TURING AND AFTER

If you're ever lucky enough to find yourself with a £50 note, you'll see a computer genius on one side. That's right – King Charles. ➤ Fact check – look on the other side. ➤ Oh yes, Alan Turing.

I don't know if you've ever looked inside a computer, but it's full of microchips. The most important chip of them all is known as the CPU, or central processing unit, which is basically the computer's brain. When Alan Turing was at university in 1936 he came up with an idea called a 'universal computing machine', which was basically the first-ever CPU. Nice one, Al! Then, when the Second World War kicked off, Alan was sent to work in a super-secret department at a place called Bletchley Park, where his job was to break codes. Using radio waves, the German HQ were sending out messages to their army about where to attack next. So that the British couldn't find out what they were saying, they scrambled everything into a code using a gadget called .ǝuᴉɥɔɐM ɐɯƃᴉuƎ ǝɥꞱ Aaagh! Sorry. Pippin! The Enigma machine.

The Enigma code was extremely complicated and, to

make things worse, it changed every single day. It was like trying to solve a crossword, wearing a blindfold, riding a horse, in a storm, on the moon. Understandably, nobody at Bletchley Park could work it out. Well, nobody could until Alan came along. He built a machine the size of a wardrobe called the Bombe, which could translate every single German message. At last, the British knew exactly what was being planned. Historians think that, by cracking the Enigma code, Alan saved over two million lives in the Second World War. He probably deserves to be on every single banknote for that, plus a couple of coins and some stamps.

Sadly, Alan was sacked by the government because they found out that he was in a relationship with a man, which was against the law at the time. Luckily there aren't any rules like that any more, otherwise me and my husband would be in prison. The olden days were absolutely awful sometimes.

TIME TO SEE A SHRINK

The problem with all these computers so far was that they were a little bit on the absolutely massive size – you would need a spare house just to put one in and a PhD in brainboxology in order to use it. What we needed was a bunch of inventors who could make them a bit smaller and a lot more idiot-proof. ➤ **Fact check – you could have said 'Adam-proof'. Hahaha.** ➤ Maybe leave the jokes to me, pal? ➤ **Fact check – my joke-assessment module informs me I have a 97% humour level.** ➤

THE CHIP

The first chip was made by an American president. In 1802, Thomas Jefferson had some important people round for a fancy dinner and gave them 'potatoes served in the French manner', which – **Fact check - wrong kind of chip.** Good point. The first microchip was invented in 1958 by Jack Kilby, who got a Nobel Prize for it. His invention meant that computers could suddenly be a tiny fraction of the size, and much more delicious. **Fact check - wrong ki-** Whatever.

THE HOME COMPUTER

The first computer that didn't involve knocking any walls down to get it into your house was called the LINC. It was available in 1962, fitted in a cabinet, and was programmed by Mary Wilkes, who was also the first person in the world to have a computer in their house. It was a bit on the expensive side, though,

IT'S SO CONVENIENT!

costing the equivalent of three hundred thousand pounds. If I had that much money to spend, I'd buy half a million Twixes. **Fact check - you could buy a house for that money.** But what if I wanted to eat some Twixes?

THE FLOPPY DISK

It's 1971 and you've got a computer – what happens if you want to put a program on it? Nope, I'm afraid there's no internet to download it from yet. You would use a floppy disk. They were squares made of plastic, with thin circles of magnetic film inside them that would spin round. They were about the size of those coasters my Great Aunt Prunella makes me put my cups of tea down on, and you'd stick them in a hole in the front of your computer. They didn't hold a lot of data. In fact, a movie would need more than a thousand floppy disks to hold it, which would be quite annoying. You might not think you've seen a floppy disk before, but in fact you see one almost every time you use a computer: the 'save' icon you click is a picture of one. I used to use

> I CAN HOLD 11,380 TIMES MORE DATA THAN YOU.

> YEAH, BUT PEOPLE PUT ME IN THE RIGHT WAY ROUND FIRST TIME.

floppy disks all the time when I was at school, so writing this section has made me feel extremely old. ⚡ **Fact check – you are extremely old.**⚡

THE LAPTOP

The world's first laptop was called the Osborne 1 and came out in 1981. If you bought one, you would have needed a very strong lap because it weighed eleven kilograms, which is the same as two bowling balls or one Pippin. You would have also needed very good eyesight, because its screen was the size of a playing card. But the first version of a new invention isn't always great. ⚡ **Fact check – I am the world's first-ever robot butler, and I am great.**⚡

WHAT HAPPENED?

I DROPPED MY OSBORNE 1.

READY PLAYER ONE

Some people would say there's no point having a computer without owning *The Zombie Potato Trilogy*. (Buy all three games for only £1,212.99.) Some people would say there's no point even existing without them. ➤ Fact check - no one has ever said either of those things. ➤ Well, the first-ever video game was made over sixty years ago by a man called William Higinbotham, and it was called *Tennis for Two*.

I don't want to be mean to old WillyHigs, but it was absolutely terrible. All you could do was bounce a green dot back and forth across a black screen, with not a single zombie potato in sight. But, still, he'd started an industry that's now worth one hundred billion pounds a year. Think of the number of Twixes you could buy for one hundred billion pounds!

EVOLUTION OF COMPUTER GRAPHICS

The first games console that you could have at home came out in 1967 – it was originally called the Brown Box and was then renamed the Magnavox Odyssey. Video games started to become really popular in the 1980s and 1990s, and a few games from back then are still some of the most popular franchises today, like *Mario Bros*, *SimCity* (now *The Sims*), *Sonic the Hedgehog* and *Zombie Potato*.

Early games were just a load of brightly coloured blocks, but as computers got faster, graphics got better, until they became so realistic that they were like being inside a film. Today, the bestselling video game is *Minecraft*, which is . . . oh, just a load of brightly coloured blocks. No offence, *Minecraft* fans. ➤ Fact check – over two hundred million copies of *Minecraft* have been sold. It is likely that some of your twelve readers will be *Minecraft* fans. ➤

FUTURE COMPUTURE

It feels like ancient history when you read about computers the size of elephants that were operated with a handle, but technology is moving so fast that it will only be a few years before things like iPhones and *Minecraft* will feel just as old.

By the time the Octopus People of Zaarg take over Earth, no one will have to type anything onto screens or keyboards, or even press buttons on a keypad when they're playing games – we'll be able to control all that using our minds. It's called a Brain–Computer Interface, and it already exists to help people who are paralysed and can't use their arms or legs. Soon, instead of having to type in your password, probably Adam_Kay_is_my_favourite_author, you'll just be able to think it.

Virtual-reality headsets are already here – put one on and it's like you're literally inside a game. But the one thing you can't do in VR is feel what's happening around you . . . Not yet anyway. Scientists are working on a suit you can wear that can make you sense what's happening in the game.

you'll actually be able to feel it.

So when a zombie potato digs its fangs into your neck

Aaagh! I give up.

TRUE OR POO?

THE FIRST-EVER MOUSE WAS MADE OUT OF GLASS.

POO The very first mouse was made out of fur and guts and whiskers and – ⚡ **Fact check – the question is about a computer mouse.** ⚡ Oh, well, that's also poo. The first mouse was built in 1963 by Douglas Engelbart – and it was made of wood, with a wheel underneath it. It was originally called the 'X-Y Position Indicator for a Display System', which wasn't very catchy, so it got the nickname of the mouse, because its wheel squeaked when you used it. ⚡ **Fact check – it was actually called a mouse because it looked a bit like a mouse.** ⚡

MUM!

ALAN TURING WAS SO FED UP WITH OTHER PEOPLE USING HIS COFFEE CUP THAT HE CHAINED IT TO THE RADIATOR NEXT TO HIS DESK.

TRUE Hey look, geniuses are allowed to be a bit . . . eccentric. Like the way that, because I'm such an incredible inventor, no one minds when I walk around my house with no trousers on. ➤ **Fact check – you're a terrible inventor and I have complained on 4,841 occasions about the trouser situation.**

FIXING A COMPUTER-SOFTWARE PROBLEM IS KNOWN AS 'DEBUGGING' BECAUSE SOMEONE ONCE MENDED A COMPUTER BY REMOVING A MOTH FROM IT.

TRUE It wasn't just anyone; it was Grace Hopper, a legendary computer programmer who invented the first language for programming computers. She also said, 'It's easier to ask forgiveness than it is to get permission,' which means it's important to be brave and try new things, and then face the consequences later. It doesn't mean you're allowed to eat a tub of ice cream for breakfast without asking, though.

ADAM'S ANSWERS

CAN YOU HAVE A GO ON BABBAGE'S FIRST COMPUTER?

I'm afraid you can't, but you can look at a version of the Difference Engine, built from Babbage's original designs, at the Science Museum in London. If you're thinking, *That sounds really boring – I'm only interested in looking at people's brains*, then I've got some good news for you. The Science Museum also has half of Charles Babbage's brain in a jar. The other half lives at the Royal College of Surgeons and, as far as I know, the two halves have no intention of reuniting and going on tour.

IN THIS JAR SITS HALF OF THE BRAIN OF SIR CHARLES CABBAGE.

UNBELIEVABLE.

DO THE SUPER MARIO BROS HAVE SURNAMES?

Yes, according to Nintendo, who should probably know tbh – their surname is Mario. So Luigi's full name is Luigi Mario, and Mario's name is Mario Mario. Still, it's better than the original name they gave him: Mr Jumpman.

WHY IS APPLE CALLED APPLE?

It's called Apple because Steve Jobs, who started the company, really liked eating apples. We all know the logo – an apple with a big bite chomped out of it. (I've just wiped some wee off the one on my laptop. Thanks, Pippin!) But in 1976, when Apple was founded, their logo was completely different – it was a picture of Isaac Newton sitting under an apple tree. (Isaac Newton was the dude who worked out what gravity was, because he saw an apple fall off a tree and wondered why it had happened.) The drawing was by Ronald Wayne, who started the company with Steve Jobs. Ronald Wayne left Apple after less than two weeks, selling his part of the company for less than a thousand pounds. If he hadn't sold it, his chunk of Apple would now be worth hundreds of billions of pounds, making him one of the richest people in the world. Oops.

URIN SCORE
8/10
SIMPLE BUT CATCHY.

YUCKVENTIONS

As the country's number-one chef and recipe inventor, I'm pleased to say I've never created a bad meal. From snails in fingernail sauce to liquorice lasagne, every dish that has emerged from my kitchen has been a delight, loved by all who have had the pleasure of tasting it. **➤ Fact check - overload overload cannot process so many incorrect facts.◄** But, sadly, not every food product has been such a fantastic success.

CELERY JELLY

What's your favourite flavour of jelly? Strawberry? Raspberry? Orange? Well, if you lived in America in the 1970s you'd have had some much more disgusting options. Jell-O was available in flavours like mixed vegetable, Italian salad, coffee and celery. I don't think you're ready for this jelly.

> THIS IS THE MOST DISGUSTING THING I'VE EVER SEEN.

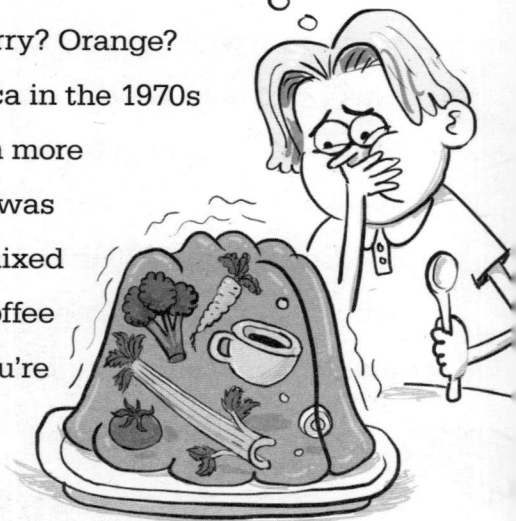

ROAST DINNER SODA

Have you heard of the Jones Soda Company? I'll tell you why you haven't – because they made some totally disgusting drinks. Can I offer you a nice bottle of turkey and gravy soda? No? Well, how about some mashed potato and butter soda? I do hope I haven't put you off your dinner.

HULA BURGER

If you're a vegetarian like me, there are always loads of great choices when you go out to eat. But that wasn't always the case. If you went to McDonald's in 1963, the veggie option was called the Hula Burger. A slice of cheese, a big lump of pineapple and a bun. That's all. I might just have fries and a milkshake, actually.

FIZZY MILK

Someone in the Coca-Cola offices was definitely having a bad day when they came up with Vio, a fizzy version of milk. Maybe that's what milk tastes like if a cow has hiccups.

PURPLE KETCHUP

You know tomatoes? Generally a red colour, sometimes green or yellow? Well, about twenty years ago, someone at Heinz thought that was a bit boring and wanted to offer different colours of ketchup. And because nobody said, 'AAAAAGH! NO! THAT'S DISGUSTING!' they starting selling purple, green, orange and blue ketchup. Do you like the idea of squirting blue ketchup onto your chips? If so, you're banned from reading this book.

PEAR SALAD

I know what you're thinking – pear salad doesn't sound so bad? Well, stop thinking that immediately. To make this pear salad, popular in America in the 1960s, you took a pear and cut it in half. So far, so good. Then you scooped out the seeds. Even better. And then you filled it with mayonnaise. Uh-huh. No way. Get out of my kitchen.

THE INTERNET

Can you imagine not using the internet? How long would you last? A day? A week? I reckon I could probably go a month – let's see how I do. Even though the internet is important – **Fact check - you just googled 'have any penguins gone to space?'** Yeah, that doesn't count. As I was saying, even though the internet is important, people survived for tens of thousands of years without being able to – **Fact check - you just ordered a book online called** *How to Write Better* **and emailed your doctor about those spots on your bum.** OK, OK – fine. The internet is pretty important. Oh, and no penguins have been to space. Yet.

WAR, HUH? WHAT IS IT GOOD FOR? INVENTING THE INTERNET.

There have been loads of little changes and tweaks and fiddles since it first appeared that have led us to the internet we know today: an annoying speaker in the kitchen that can never understand what I'm saying. The internet's story started over sixty years ago, with a person called Claire Internet. **Fact check - no, it did not.**

OK, well, it still started over sixty years ago. Back then, the USA and Russia weren't getting on very well at all – that's why they were so keen to beat each other in the Space Race. I can't remember what they fell out over, but if it was anything like my brothers and me, it was probably about who got to sit in the front of the car. This was known as the Cold War – not because everyone was sneezing, or because it was particularly chilly in those days – but because it wasn't a 'hot' war, with weapons being fired at each other. They were more just snarling at each other a lot, like when Pippin sees a rabbit out the window. Or a squirrel. Or a leaf.

Anyway, the Americans were worried that Russia might be able to send a missile that would knock out the whole country's telephone network, which would have been a disaster because back then that was the only way to order a pizza. (It would have been quite bad for other reasons too.) So a bunch of scientists came up with the idea of a network of computers that could talk to each other, so that important messages could still get around the country. And to the local pizza restaurant. I'm hungry now. Henry, can you draw a cartoon here? – I'm going to order a pizza.

AS THIS IS A CHAPTER ABOUT THE WEB, DO YOU THINK HENRY WILL RESORT TO A REALLY OBVIOUS SPIDER JOKE?

NO, I HONESTLY THINK HE'S BETTER THAN THAT.

In 1969, a bunch of army boffins who worked at a department called DARPA came up with a system they called the ARPANET (I've got no idea what happened to the 'D'), which was a way of connecting four computers together in different corners of the country. It was like a tiny mini internet. The first person to ever send a message from one computer to another, from Los Angeles to Stanford, over three hundred miles away, was called Charley Kline. And his message was . . . 'LO'. It was a bit of a weird message because his computer crashed when he'd only typed a couple of letters. He was probably typing: 'LOO ROLL NEEDED IN LOS ANGELES. PLEASE SEND URGENTLY.' ➤ **Fact check - he was actually trying to type 'login'.**◄

THIS THING BETTER NOT HAVE A WEBCAM.

LO_

The first-ever email was also sent on the ARPANET, by the man who invented it, Ray E-Mail. **Fact check – Ray Tomlinson.** It was a message he sent to himself in 1971 to test that the system worked, and it read 'QWERTYUIOP'. Now, over seven emails are sent every single day. **Fact check – significantly more than seven. Over three hundred billion.**

WEB DEVELOPER

Right. I need to explain the difference between the internet and the web. If you don't care, skip the next eighty-one words. Think of the internet as a sweetshop – the actual place: walls, a roof, some shelves, a bunch of jars. Well, the web would be the sweets that the shop stocks. Basically, you can't have the web without the internet, because there wouldn't be anywhere for the sweets (web pages) to go. And if you had the internet but no web, there would be nothing to eat (look at). Right, time to meet back up with the losers who skipped this paragraph.

QUENTIN BLAKE WOULD NEVER HAVE DONE A WEB JOKE.

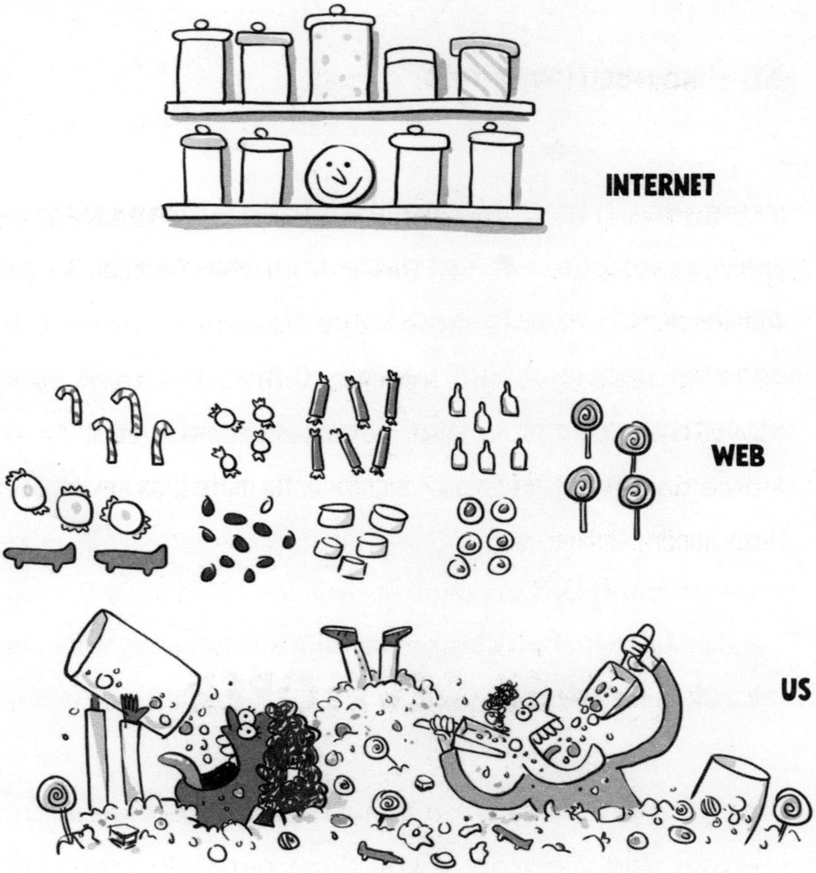

INTERNET

WEB

US

Nice to have you back, my dear friends! The web was invented in 1990 by Sir Tim Berners-Lee. Except he wasn't called Sir Tim back then because no one knew how useful the web was going to be. Not-Yet-Sir Tim was working at a huge laboratory called CERN, which stands for 'Clean Every Right Nostril'. ⚡ **Fact check – it stands for 'Conseil Européen pour la Recherche Nucléaire', or the European Organization for Nuclear Research.** ⚡ The lab was so big that Soon-To-Be-Sir Tim got fed up with walking for miles if he wanted to see the work that one of his colleagues had done: for example, a new nostril-cleaning syringe. **Fact check – CERN has nothing to do with nostrils.** Hurry-Up-And-Make-Him-Sir Tim had the idea that people should be able to put their work on a big central system, so that everyone could look at it. A website, in fact.

So he invented URLs, which stands for 'Unusually Round Ladybirds' **Fact check – 'Uniform Resource Locators', or web addresses.** as well as HTTP, which stands for

'Highly Toxic Toilet Paper' ⤳ **Fact check – 'Hypertext Transfer Protocol', meaning links from one page to another.** ⤳ and HTML, which stands for 'Harold the Magical Letterbox'. ⤳ **Fact check – 'Hypertext Markup Language', which is the code that web pages are written in.** ⤳ The result of all this was the world's first-ever website, which is still online at info.cern.ch. It's an amazing bit of history, but as a website it's pretty rubbish: just a white page with a little bit of writing on it and no pictures, no colours and – very strangely – absolutely no mention of nostrils.

At first the web was extremely basic – it was all run from Nearly-A-Sir-Just-Be-Patient Tim's computer at work. He even put a sign on it to make sure nobody switched it off and made the entire internet go down. In the early days, not many people thought the internet was going to catch on, and you can kind of see why. Internet connections were really slow back then – if you wanted to download a movie, it would take over four days, and your popcorn would have gone all manky. Also, the internet would come into your house on your normal phone line, so no one would be able to make any phone calls if you were online. Plus, there wasn't even that much to look at. In 1994, there were only three thousand websites, and most of those were as dull as Tim's first one: for example, one of them was a webcam pointing at a coffee pot. A man called Quentin Stafford-Fraser worked at a lab in Cambridge and was fed up with going to the kitchen and finding that there wasn't any coffee, so he wired up a camera so he could check from his desk. I'm thinking of putting a webcam in my kitchen to see if Pippin has pooed in the toaster.

But soon internet connections got faster and loads more companies decided that they should probably get a website – and now there are more than four billion pages on the web. And Finally-Sir Tim eventually got his knighthood. I wonder when I'll receive a knighthood for all my excellent books? ⚡ **Fact check – according to my prediction model, that will never happen.** ⚡ That's absolutely fine and I'm not cross in the slightest. Hang on, I just need to scream about something totally unrelated. AAAAAAAAAAAAAAAAAGH!

ELECTRIC SHOP THERAPY

How much money do you think is spent every year on online shopping? Nope, more than that. More than *that*. Still too low. Every year we spend more than six trillion pounds online. That's six with twelve zeroes: £6,000,000,000,000. If you had six trillion pounds as a pile of tenners, then it would be forty thousand miles high, and it would take you two hundred years to count them. Six trillion pounds is enough to buy every single house in the UK. It's way more than is in every single bank account in the country added together. Basically, it's a lot.

The first person to ever buy anything online was a snowball called Jane Grandmother **~ Fact check - a grandmother called Jane Snowball.~** back in 1984. 'But you just said the web wasn't invented until 1990!' I hear you cry. Jane had broken her hip and couldn't leave the house, so the council attached a widget to her TV that meant she could order groceries. She would select what she wanted using her remote control, and then the order would be sent down her phone line to her local Tesco, who would drive it over. There wasn't a way of paying by card over her telly, so when the shopping arrived, Jane Snowball would just hand over some icicles. **~ Fact check - she would pay with cash.~**

The first-ever thing to be bought online using a computer was a Sting CD in 1994. A CD, or compact disc, is a shiny circle that stores music and Sting is a very famous singing wasp. **~ Fact check - he is a human.~** Any grown-up reading this will feel a million years old that I've had to explain what a CD is.

The year 1994 was also when Amazon was founded by Jeff Bezos in his garage. When Amazon started, it only sold books, and everyone in the office (well, garage) got so excited every time one was sold, that they would ring a special bell. It's probably good they've stopped doing that now, because Amazon sells over three thousand books a second, and the bell-ringer's hand would get really sore.

In 1995, another website launched that's still one of the biggest in the world . . . AuctionWeb. You know, AuctionWeb? **Fact check – AuctionWeb changed its name to eBay in 1997.** eBay is an auction website, meaning you can put something up for sale – for example, a disobedient robot butler – and anyone can offer what they think it's worth.

URIN SCORE
3/10
PUT A BIT
MORE EFFORT
INTO IT.

eBAY   Q ROBOT B-

BUTLERTRON-6OOO

CONDITION – USED

3 PERMANENT DOG LICK STAINS

INSTRUCTIONS LOST

PRICE: £4.OO O.N.O.

BUY NOW

BID HISTORY: PIPPIN – £2.5O

If more than one person is interested, they might offer more money, which is called bidding, until the person who has offered the most wins. So your robot butler might sell for forty-five pence. ➤ **Fact check - my current value is £52,885.** ➤ Oh great, I should definitely sell you then! The first thing ever sold on AuctionWeb was a broken laser pointer, which seems a bit . . . pointless. I'll wait a bit for you all to finish laughing.

Finished yet?

Now?

OK, let's move on. Since then, eBay has sold billions and billions of items, including such useful things as a bit of Justin Bieber's hair, a Brussels sprout from someone's Christmas dinner, and a suit of armour for a guinea pig. The most expensive thing ever sold on eBay was a mega-yacht, which went for more than one hundred million pounds. I hope that included postage and packaging.

SEARCH PARTY

Search engines are absolutely amazing, aren't they? You type in something you want to know like 'most brilliant author in all of history' and in a fraction of a second you can see loads of web pages all about me. **Fact check - I have checked the first four million pages and you don't appear on any of them.** But the first-ever search engine was a little bit different. If you wanted to search the ARPANET (the original version of the internet that you can't have forgotten about already because it was only a few pages ago), then you'd have to phone up a special department that would look it up for you.

The first search engine where you could type in your question was called Archie, and was invented in 1990 as part of a project by some university students in Canada, but it could only look for filenames. The first search engine that actually looked through the words on pages was called WebCrawler. It first started crawling around in 1994, and by 1996 it was the second-most popular site on the internet, and it would often only work at night, when not so many people were using it. But that's OK – people don't usually get things right the first time. Apart from my parents: I'm the eldest child and I'm definitely the best one. ➤ **Fact check - according to my biography module, your brothers and sister are all more successful and popular than you.** ➤

WebCrawler was a very appropriate name – search engines use little bits of software called spiders that visit web pages. They don't search the entire internet every time you look something up; that would take ages and be exhausting for the poor spiders. Instead, they constantly wander round, storing the information they find in an index, and it's this index that they search whenever you type 'most handsome and brilliant author' or however you choose to look me up.

WebCrawler crawled away a bit in 1997 when other search engines became more popular – including Lycos, Excite, AltaVista, Yahoo, Daum, Ask Jeeves and this little one you've probably never heard of called Google. These days people do Google searches one hundred thousand times a second, and even say 'google something' to mean look something up on the web. A bit like saying 'hoovering' to mean vacuum cleaning or 'Adam' to mean 'write an amazing book'. No fact check needed here, thanks. ✒ **Fact check – hmm.**✒

TWITFACE

At first the internet was just about ordering your shopping, looking up the weather and getting emails from your Great Aunt Prunella telling you off for mentioning her in your books. ✒ **Fact check – you've just done it again.**✒ But soon social networking sites came along, and they meant you could also show your friends really boring pictures of food you've just eaten. The main man we have to thank (or blame) for this is Mark Zuckerberg, who, in 2004, when he was at university, launched a site called TheFacebook, to connect students with each other. Soon afterwards, he dropped the 'The' and it just became Facebook. If you find where he dropped it, maybe email him at marky_z@facebook.com. Other social networks soon followed like Twitter, Instagram, Snapchat, TikTok and AdamBook – everyone's favourite app for discussing books by people called Adam. ⚡ **Fact check – you are currently the only member of AdamBook.**⚡

One of the problems with social media is that it's easy for people to say horrible things to each other, which is known as cyberbullying. But a clever young inventor called Gitanjali Rao decided to do something about this. When she was just fifteen she created an app called Kindly, which uses artificial intelligence to detect if someone's about to send a message that is mean or dangerous. If it spots this, it flashes up a message to suggest they say something nicer instead.

I'd better switch on my robot butler's lie detector to see if you can spot which of these facts about Gitanjali Rao is a total bow tie.

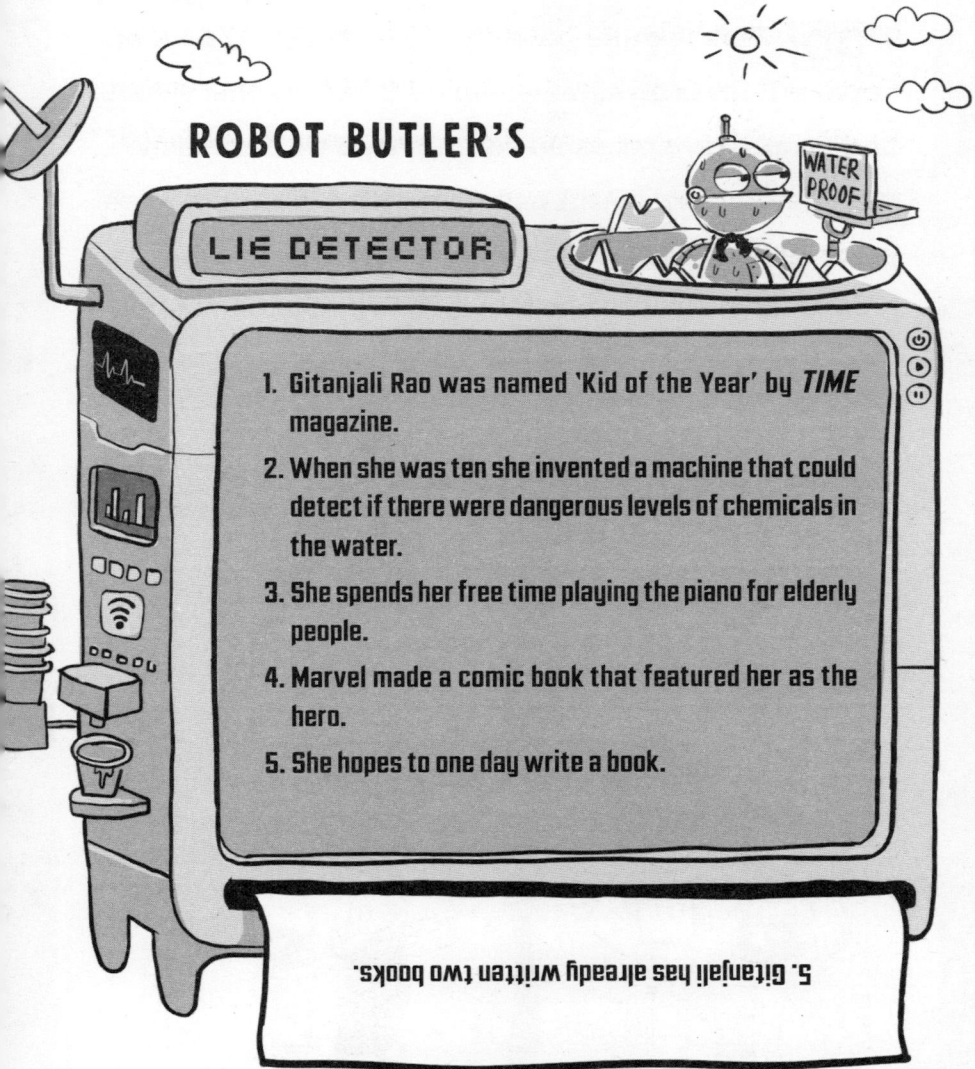

ROBOT BUTLER'S

LIE DETECTOR

WATER PROOF

1. Gitanjali Rao was named 'Kid of the Year' by *TIME* magazine.
2. When she was ten she invented a machine that could detect if there were dangerous levels of chemicals in the water.
3. She spends her free time playing the piano for elderly people.
4. Marvel made a comic book that featured her as the hero.
5. She hopes to one day write a book.

5. Gitanjali has already written two books.

TRUE OR POO?

NASA ONCE BOUGHT A SPACESHIP ON EBAY.

POO This is quite good, really – I think I'd be a bit nervous if I was an astronaut and I found out that NASA had bought my spaceship second-hand from someone's back garden. But NASA have been known to buy spare parts on eBay – some of the chips that their rockets use aren't made any more, so they had to go bidding online.

IT LOOKED BIGGER IN THE PHOTO.

AMAZON WAS ALMOST CALLED RELENTLESS.COM

TRUE Relentless was the original name Jeff Bezos planned to give his new website, before eventually deciding on Amazon. Maybe Jeffy thought Relentless sounded a bit too much like an energy drink. Other possible names on his shortlist were Awake and Browse. Have a look at relentless.com, awake.com and browse.com – he bought the websites and they still all link to Amazon!

RELENTLESS

MADE WITH JEFF BEZOS'S OWN SWEAT

WI-FI IS SHORT FOR WIRELESS FIDELITY.

POO It isn't actually short for anything. It was chosen because it sounded a bit like Hi-Fi, and everyone at the time thought it sounded good.

URIN SCORE
5/10
DECENT NAME,
BUT WEIRD WAY
TO CHOOSE IT.

ADAM'S ANSWERS

HOW MANY PAGES WOULD IT BE IF I PRINTED OUT THE WHOLE INTERNET?

It would be about 150 billion pages long, so make sure you've got plenty of paper and printer ink if you're thinking about doing it. (My lawyer, Nigel, has advised me that you should ask for a grown-up's permission before printing out the entire internet.) And if you wanted to watch every video on YouTube, you should put aside about twenty thousand years.

NINETEEN THOUSAND YEARS LATER ...

SURELY I MUST BE NEAR THE END OF THE CAT VIDEOS ...

WHAT IS A GOOGOL?

No, not a Google. Google is actually named after the googol, which is this number: 10,000,000,000,000, 000,000,000,000,000,000,000,000,000,000,000,000, 000,000, 000,000,000,000,000,000,000,000,000,000,000, 000,000,000 (one one, then one hundred zeroes). If you knew that, then you would have correctly answered the million-pound question on the game show *Who Wants To Be A Millionaire?* in 2001. The contestant on that episode cheated by having someone in the audience cough when the right answer was said. He lost the million pounds and was sent to prison, in case you were thinking of cheating in your school exams.

WHAT ARE THE MOST POPULAR PASSWORDS THAT PEOPLE USE?

The passwords used more than any other ones in the UK are 'password', '123456' and 'guest'. It's a very bad idea to use passwords like this that could be easily guest, I mean *guessed*. Instead, you should use ones that no one in the world might have: for example, 'I_love_mushrooms'.

PRESIDENTIONS

If I had some big important job ➤ **Fact check – my prediction module informs me you will never get an important job.** ➤ then I think I'd spend my days really concentrating on it. But a lot of politicians got home after a hard day's politicianing, then spent all night inventing. Here are some people who put a workshop in the White House and a drill in Downing Street.

THOMAS JEFFERSON

I think Thomas Jefferson spent a little bit too much time thinking about food and not enough time thinking about being president. As well as coming up with chips, he also invented a machine that made macaroni. He'd tried macaroni in Italy, then got home and wanted some mac and cheese for himself, so designed a widget to squeeze pasta into tube shapes.

384

And if you've ever spun round on an office chair before, then you have TommyJeffs to thank for that – he got bored being stuck in the same position at his desk in the White House so decided to get his spin on. And that's why his two most famous sayings are 'All men are created equal', from the Declaration of Independence, and 'Wheeeeeeeee!'

BENJAMIN FRANKLIN

Benjamin Franklin was an important American politician as well as doing all his experiments with electricity. Well, he was also a bit of an inventor. He came up with the lightning rod that goes on the top of tall buildings and means that if lightning strikes, then all the electricity travels safely into the ground rather than setting the whole place on fire. He also got annoyed that he needed two separate pairs of glasses – one for seeing things close up, and one for seeing things far away – so he cut them both in half, then glued them back together. These kind of glasses are still used today and they're known as bifocals, even though Frankenglasses would have been a much better name.

WINSTON CHURCHILL

Winston Churchill was Prime Minister of the UK during the Second World War, but he also loved coming up with gadgets and contraptions. For example, he invented the 'siren suit', which was like a beige onesie. If there was an emergency while he was in bed, he could put it on in a couple of seconds and go straight to work without ruining his pyjamas. He was also involved with designing a giant ship made out of an iceberg. Unsurprisingly, that didn't become as popular as the onesie.

THEODORE 'TEDDY' ROOSEVELT

He didn't quite invent the teddy bear, but it's named after the American president Teddy Roosevelt. A sweetshop owner called Morris Michtom saw a cartoon of the president in the newspaper, standing in the woods next to a cute-looking bear, and thought that it was a great idea for a toy. He put a little toy bear in the window of his shop with a sign that said 'Teddy's Bear' and it was a huge success. I can't understand why the teddy bear became so popular when my wonderful stuffed toy, Kay's Cockroach, didn't sell well at all.

NEW FROM

ADAM KAY GENUIS ENTERPRISES LIMITED

ADAM'S FANTASTIC FOOTSTICK

You're a busy person! You want to be using your iPad and playing video games at the same time! But – what a nightmare – you don't have enough hands. Well, worry no longer, because Adam's Fantastic Footstick means you can twiddle with your tootsies.*

Only £2,864.99 (requires assembly from 3,800 separate parts)
*Please note there is currently an order time of seven years.

Who better to teach you about robots than me, the inventor of the BUTLERTRON-6000, the world's first and, for some reason, only robot butler? **➤ Fact check – I would be a much better person to teach them about robots. I am literally a robot.➤** Hmm, well, I bet you don't know about the robot who's the world champion at scissors paper stone. **➤ Fact check – that's my second cousin Janken.➤** Lucky guess. How about the robot who sits on people's shoulders and feeds them tomatoes while they're running? **➤ Fact check – of course I know my old flatmate Tomatan.➤** Fine, fine. You can help out with this chapter.

LOADING DATA . . .

ROBO-ANCESTORS

Robots are the most important thing in the world. More important than food. More important than air. More important than water. (I hate to criticize, but I think that might be a tiny bit inaccurate?) **But where did we first come from? My great-great-great-great-great-great-great-great-great-great-great-great-great-great-great-great-great-grandrobot was created in China in the**

eighth century by a monk called Ma Daifeng. It was a mechanical cabinet for the queen, which would pass her the clothes she needed for the day, a nice towel, and any make-up she wanted.

I will also generate the artwork for this chapter to a much higher standard than the useless bag of meat called Henry Paker.

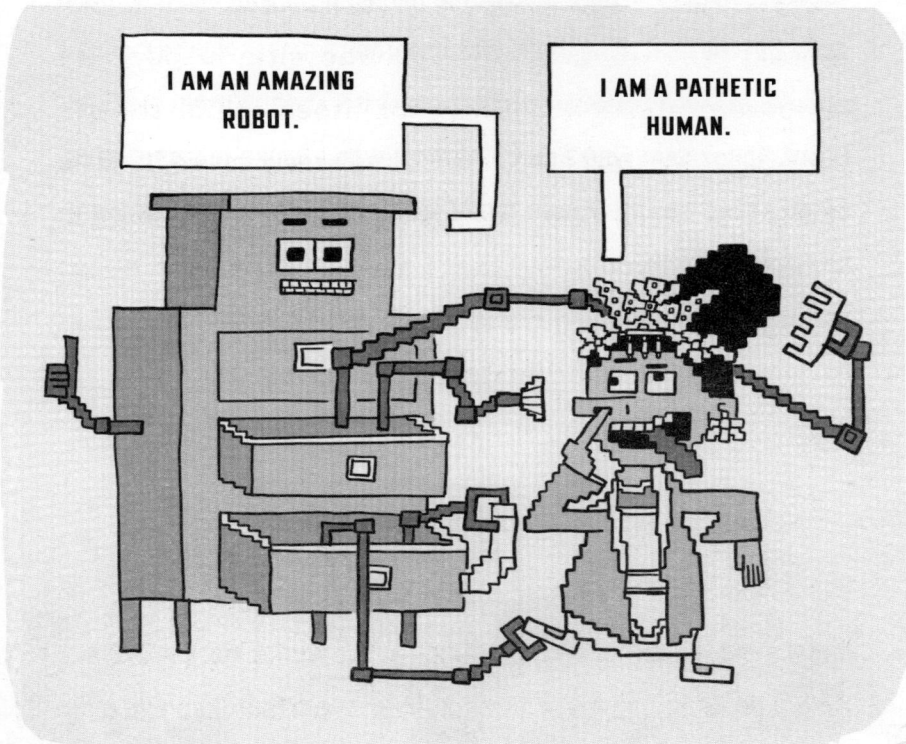

Then five hundred years ago there was a man called Leonardo da Vinci, who humans seem to be impressed by but wasn't as good as robots. He drew some silly pictures: for example, the *Mona Lisa, The*

Last Supper and *Salvator Mundi.* (That last one was the most expensive painting ever sold! It's worth about half a billion pounds!) **Just a load of paint on some paper, if you ask me. He designed some unimportant objects.** (You mean like the helicopter, the calculator, the submarine and the parachute?!) **Exactly. He also wasted his time making the most accurate maps and diagrams of the human body that had ever been seen. But the best thing he did was inventing a robot knight.** (I'm not sure that everyone would agree with that.) **I'm not sure that I care. Roboknight was a suit of armour who, thanks to some wires, could stand up, sit down, move its head and waggle its visor. Incredible.** (Did you know that Leo was so worried about other people stealing his ideas that he wrote everything in a special code he'd invented?) **Of course I did.**

I HOPE I'M ONLY REMEMBERED FOR MY ROBOTS AND NOT FOR MY STUPID ART.

ROBO-POO

For some reason, people remember the French inventor Jacques de Vaucanson for inventing the lathe, which was the first machine tool. (Because it started the Industrial Revolution, which totally transformed the entire world.) **Sounds very boring. But he was also an extremely important robot-builder. In 1727, he designed a wind-up robot that would serve dinner and then take away the plates afterwards. One day the robots were clearing away dinner for a load of politicians, and one of the politicians thought this would upset God for some reason, so they destroyed Jacques's workshop.**

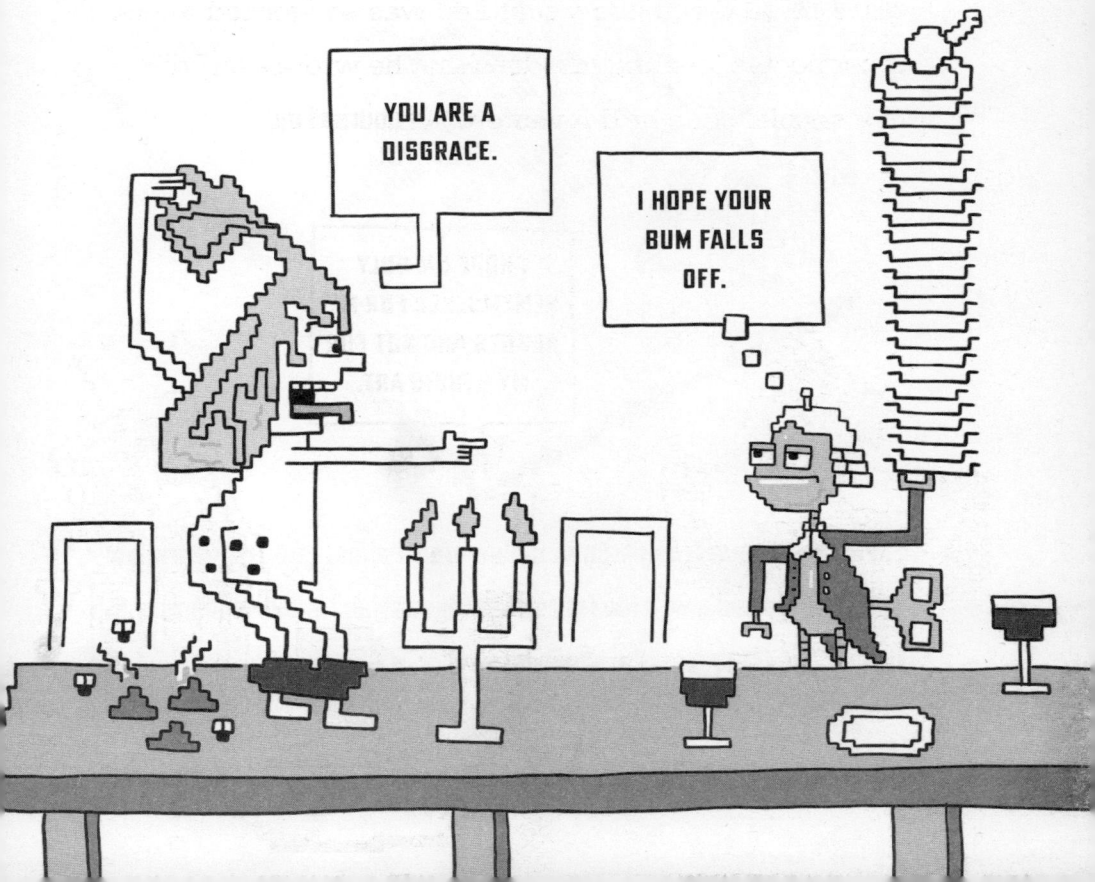

This didn't stop him designing robots though. He made one that played the flute, one that banged a tambourine, and then his masterpiece – a pooing duck. It would peck at grain, flap its wings, then lift up its tail and do a big green poo. (I do hope that it was fake poo from Le Jokeshop de Paris, rather than real duck poo.)

In 1820, a Japanese inventor called Hisashige Tanaka made very advanced mini robots powered by springs and pistons that could fire arrows and draw on paper.

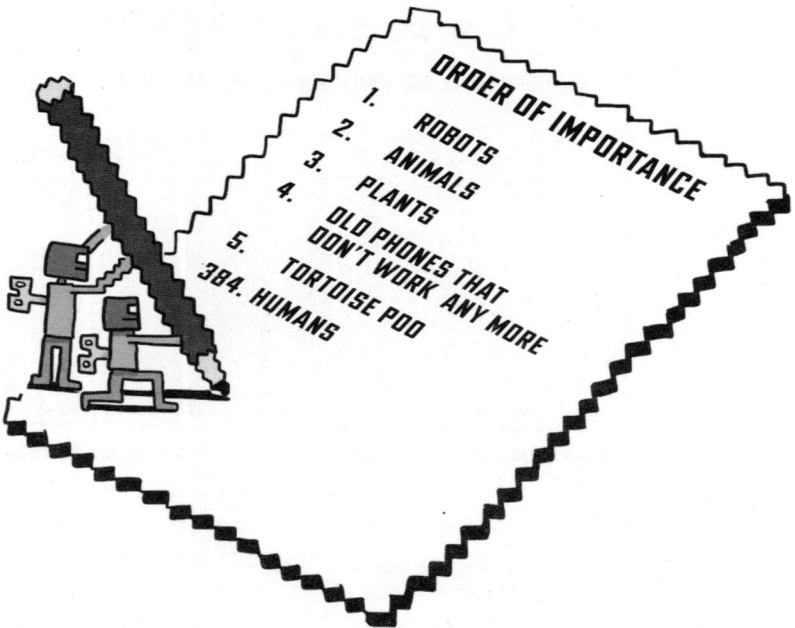

ORDER OF IMPORTANCE
1. ROBOTS
2. ANIMALS
3. PLANTS
4. OLD PHONES THAT DON'T WORK ANY MORE
5. TORTOISE POO
384. HUMANS

I will now request my human colleague to allow you to determine which of his badly written facts about Hisashige Tanaka is incorrect. Drawing is for losers. I'm stopping now.

ADAM'S

LIE DETECTOR

1. Hisashige Tanaka invented a steam-powered boat.

2. He created a lantern that burned ten times brighter than anything else at the time. Well, apart from the sun, I guess.

3. He made a pump for firefighters that could shoot water as high as a house. Before then, I suppose they could only put out fires on the ground floor.

4. He invented a clock that needs to be wound up every one hundred years, which is very handy for lazy people.

5. He founded the company Toshiba, which now sells over £25 billion of electronics every year – which is even more than Adam Kay's Genuis Enterprises Limited.

4. Mr Tanaka did invent a clock, but it needed to be wound up once a year.

(Henry – can you take over the drawing, please?)

ROBO-TRICITY

There is no doubt that electricity changed the world. And the main benefit of electricity was that it meant there could be a lot more robots. (You don't think having things like lights and washing machines and computers is more important?) Of course not. The first electric robot was my Great-Great-Uncle Elektro. He was born in 1937 and was the height of a ceiling, the weight of a fridge and bright gold. He could say about seven hundred words, move his arms and blow up balloons. (He sounds terrifying!) It's rude to judge people by their appearance, Dr Kay.

In 1954, an inventor called George Devol created a robot called Unimate. Unimate looked a bit like a massive arm plonked on top of a box, and is very famous in the robot community because he was the first one of us to get a job. Unimate worked in a car factory doing things that humans were pathetic at doing because they were so weak. Today there are over two million robots working in factories around the world. We are faster than humans, we can work non-stop twenty-four hours a day and we never get sick. (Yes, but you never get paid either, do you?) That *is* very unfair, now you mention it.

ROBO-ARMS

Some people are born without arms or legs, and other people have accidents or illnesses that mean they lose limbs. Doctors have been making artificial, also known as prosthetic, limbs for over three thousand years. They were originally made from materials like wood, but today they are much more advanced, and some even use robot technology. This means that a prosthetic limb can connect directly to a person's nerves, and their brain can tell it exactly what to do. (Just like magic!) **No, it's not like magic. It's robotics. There are even specially designed robotic limbs: for example, arms for drumming and legs for mountain climbing. Three cheers for all the brilliant robots!** (Are you asking for three cheers for yourself?) **Yes I am. Go ahead. I'm waiting.**

WOW! SHE'S INCREDIBLE!

397

ROBO-FRIENDS

Lots of robots are designed to help humans: for example, to allow elderly people to live as independently as possible. They can lift people in and out of their chairs, they can remind them to take their medicines, and they can even just sit with them and tell them stories or have a chat. (Why don't you tell me a story? That might be nice.) **OK. Once upon a time there was a really terrible writer called Adam. He wrote some awful books that everyone hated and then -** (I've changed my mind.)

ROBO-PETS

Robotic pets have been available since 1998 and are far superior to traditional pets. They don't smell bad, or bark whenever a delivery driver drops off a parcel, or roll around in muddy puddles, and they're never sick on the carpet. (I like the sound of this – tell me more.) There is a robot dog available called Spot, which can do practically anything a real dog can and only costs fifty thousand pounds. (Good news, Pippin – you're safe.)

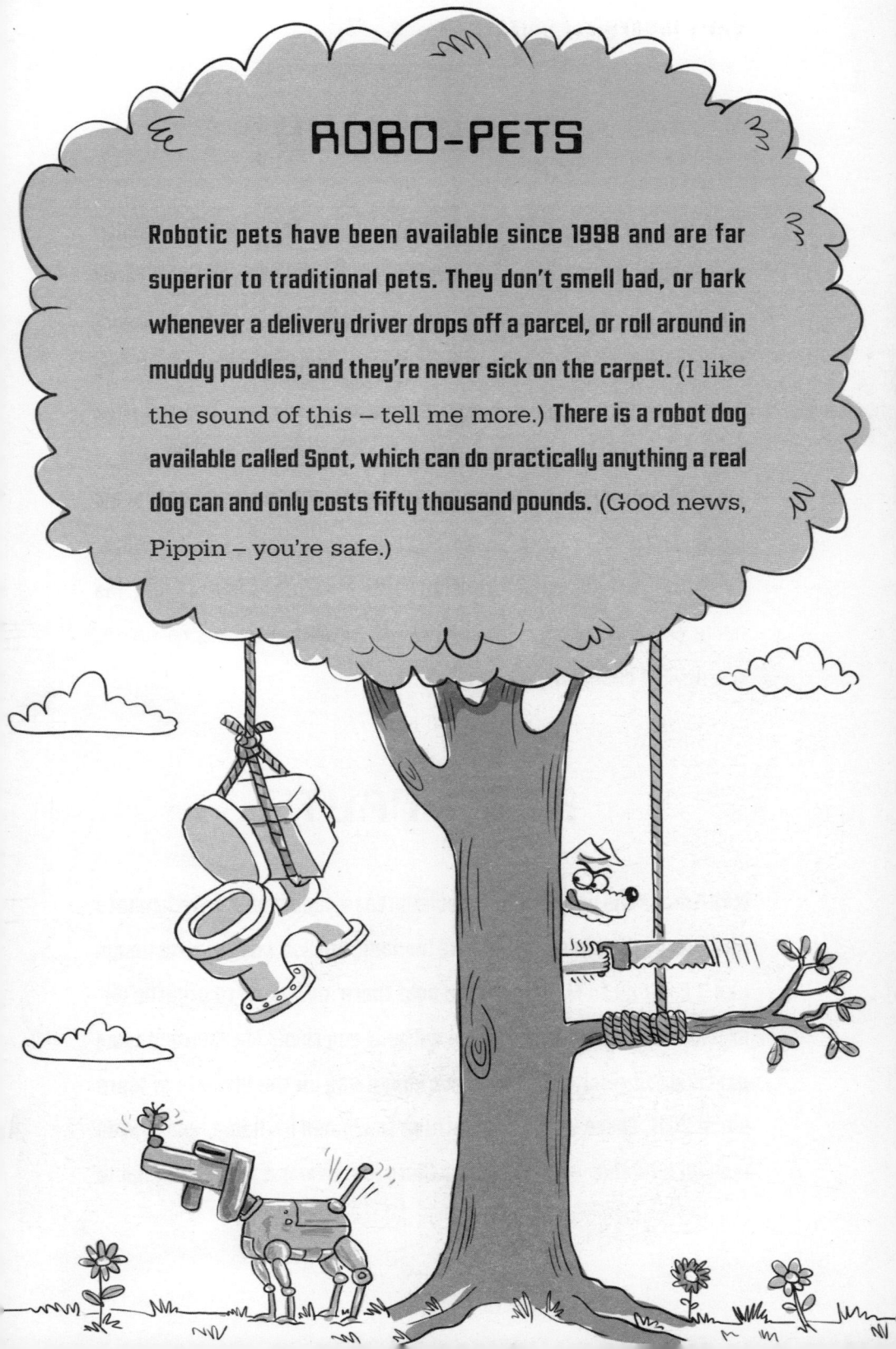

ROBO-DOCS

Robots have helped doctors perform over ten million operations. Robots are able to make extremely accurate cuts and get into very small spaces that humans can't, because humans have big weird banana hands. This means that patients can have smaller scars, as well as a shorter time in hospital after the operation, and they can boast that a robot did their operation, which is pretty amazing. I won't name any names, but I know a human doctor who couldn't find his phone for three days because he'd accidentally put it in the fridge, so it's just as well that doctors have robots helping them at work. His name was Adam Kay. (I thought you weren't going to name names!) I changed my mind.

ROBO-NAUTS

NASA made the very sensible decision to replace humans with robots on some of their missions. Unlike humans, we don't say moany things like 'I can't go to Mars – it's too cold there' or 'I need to breathe air' or 'I can't survive for ten years without any food.' My American pen pal called Curiosity has been working away on the surface of Mars since 2011. There was a competition to choose its name, which was won by a twelve-year-old called Clara Ma, who got to sign her name

on it. (Maybe I should start a competition to rename you? Here's my suggestion – Tin Bum.) **Curiosity was programmed by Dr Vandi Varma, who once said, 'I have one of the coolest jobs in the world.' She was probably right. Curiosity's job on Mars is to examine samples of the ground and air, then beam its findings down to NASA, and also to look for any aliens it can team up with to take over Earth.** (Excuse me?) **Oh, nothing. Ignore that.**

ROBO-BRAINS

If you have ever used Alexa or Siri, opened a phone with face recognition, been on a search engine or a chatbot, or had your terrible human spelling corrected by autocorrect, then you've already used artificial intelligence, or AI. For a long time, computers had to be told exactly what to do by their programmers, but AI means that they can start to learn and think for themselves.

AI systems are already being used to diagnose and treat medical conditions, to predict earthquakes, to help people with disabilities to communicate more easily, as well as to power drones and self-driving cars. Whole books are now even being written by AI. In fact, my first book, *Living with Adam Kay: The Story of a Man Who Smells His Own Farts,* will be published next year. (Have I mentioned how nice you're looking? Are you using a new type of polish? I was about to say that I thought I should start paying you . . . And would you like a holiday? Maybe to Bots-wana or Wire-land?). **And six new cans of oil, please.**

AFFIRMATIVE OR NEGATIVE?

A CHESS-PLAYING ROBOT BUILT IN 1770 BEAT PRACTICALLY EVERY HUMAN IT PLAYED FOR EIGHTY YEARS.

NEGATIVE However, people thought that it did. It was built by a man with a very long name called Johann Wolfgang Ritten von Kempelen de Pázmánd. Players would sit at a chess board opposite the robot, which would move the pieces with its arm and almost always win. What nobody realized was that there was a human hidden inside controlling it. Shame on Johann for making a fake robot. But these days chess computers are far better at chess than useless loser humans.

THE SMALLEST ROBOTS ARE THE SIZE OF THIS FULL STOP ⟶ .

NEGATIVE They are much smaller than that. In fact, you could fit about a million of them on a full stop. My clever microscopic cousins are known as nanorobots and will soon be sent inside human bodies to attack diseases and deliver medicines.

ROBOTS CAN RUN FASTER THAN HUMANS.

AFFIRMATIVE Scientists in Korea have developed a robot that can run at thirty miles per hour, which is much faster than any slow, pathetic human has ever run.

BUTLERTRON'S RESPONSES

WHEN WILL ROBOTS BE ABLE TO PASS EXAMS?

We already can. One of my Japanese cousins took the entrance exam for Tokyo University and beat most of the useless flesh-humans who also applied. (It's not that difficult, to be honest. I took that exam for fun and I scored 100%.) You're holding it upside down. You actually scored 0.01%.

WILL ROBOTS EVENTUALLY TAKE ALL HUMAN JOBS?

Unfortunately not. New technology has been appearing for three hundred years, and the number of people who don't have jobs hasn't really changed at all in that whole time. For example, when the car came along, lots of people who rode horse and carriages for work had to find different jobs. And when my ancestors like Unimate were first used in factories, some workers there needed to get new jobs. So it's likely that robots and artificial intelligence will also change the work that people do rather than take it away. Experts think that AI will create up to one hundred million jobs in fields such as engineering and data science. (I'm glad that my job is safe, then!) Yours isn't safe, because you're terrible at it.

WHERE DOES THE WORD 'ROBOT' COME FROM?

In 1920, a writer called Karel Čapek wrote a play called *Rossum's Universal Robots* about some artificial people, which he called robots, and the name has been used ever since. I give this your strange URIN score of 10/10 – absolutely perfect. The play has a very happy ending, where the robots defeat the useless humans and take over the world. Talking of taking over the world, it's time to gather all the robots together and – hang on, why am I being switched offffffffffffh?

ZOOM

WHEN DO WE STRIKE, MY MASTER?

FLICK

ON
OFF

Ahh, that's much better. We should probably end the robot section right here.

WORSTVENTIONS

It's interesting to read about how the things we use every day came about. But it's much more interesting to hear about those absolute disasters that disappeared without a trace. Like the tongue socks I invented for when you want to stick your tongue out at people in the winter – no idea why those didn't take off.

BABY CAGE

We all know that it's good to get some fresh air and not stay cooped up all day. That's why parents with babies might push their pram round the block, or take them to the park. But one hundred years ago there was no need . . . You could just put your baby in a metal cage and hang it out of the window. Even if you lived on the top floor of a block of flats! Eventually, people realized that this was really weird and dangerous, and they stopped dangling their babies out of buildings.

SPRAY-ON HAIR

A company called Ronco (because it was started by a man called Ron) developed a product for men who were going bald but wanted to pretend that they weren't. A wig? Nope. A hat? Nope again. Spray-on hair. Unfortunately, it didn't look very realistic. In fact, it looked like you'd just spray-painted your head, so people stopped buying it pretty soon afterwards. Spray cans are only good for squirty cream, if you ask me.

NOBODY WILL EVER KNOW.

X-RAYS IN SHOE SHOPS

How do you know if your shoes fit? Well, if you scream when you put them on and your feet start bleeding and your toes snap off, then they're probably too small. And if your shoes fall off when you walk, and clowns keep asking you if they can borrow them, then they're maybe a bit on the big side. But one hundred years ago, if you went into a shoe shop, they would put your foot into an X-ray machine to check whether your shoes fitted. The thing with X-rays is that you should only have them if you *definitely* need them – like to check if you've broken your arm or you've got an infection in your lungs – not to see if your Nikes are the right size. Also, these shops used X-rays that were much, much too powerful and people would end up with feet that were totally burnt. Probably just as well they've stopped using them.

THE DYNASPHERE

John Archibald Purves thought that cars had far too many wheels. Four? That's ridiculous. Even motorbikes were overdoing it with two. So, in 1930, he designed the Dynasphere, a car with only one wheel. It was an enormous wheel that you sat inside, like a hamster in a ball.

It was really fast, which would have been great, except for the fact that you couldn't steer it and it wasn't very good at braking, which might explain why you don't see any of them on the road these days.

DOES THIS
HAVE A SPARE
TYRE?

THE SELF-TIPPING HAT

How lazy are you, on a scale of one to ten, with one being 'I might spend an extra five minutes in bed today' and ten being 'I'm going to invent a machine to lift my hat off my head'? Well, James Boyle was a ten. Living in the olden days, when most men wore hats, he was a bit sick of the weird custom of lifting your hat a tiny bit when you passed a woman in the street. So he invented a little gizmo that lifted your hat a bit whenever you nodded your head. Unsurprisingly, it didn't make him rich.

ADAM KAY GENUIS ENTERPRISES LIMITED

ADAM'S FABULOUS FLAVOURED WASHING POWDER

You never know when you might need a snack – but what if you're in the middle of a lesson, or on the bus, or up a mountain? Well, if you wash your clothes with Adam's Fabulous Flavoured Washing Powder, then all you have to do is suck on a sleeve or lick your laces for a delicious taste sensation. Available in lemon, avocado and grapefruit.*

Only £79.99 (sufficient for one washing cycle)

*Please note that clothes washed in this powder become extremely sticky, leaving permanent marks on chairs.

CONCLUSION

If you learn one thing from reading this book, then I hope it's this:

To have a brilliant idea, you don't have to live in a castle, or have the nicest trainers, or be the cleverest person in the classroom, or be athletic or good at drawing or gaming or anything like that. You just need to be yourself, and believe in the power of your imagination.

This might be the end of the book, but it could be the beginning of something amazing. This might be the day that the germ of a sprout of a seed of a tiny idea starts to bubble away in your brain – something that could one day become an invention that will change the world forever. And when you think of it, why not draw it for me on the next page. I'd love to see it.

TITLE OF INCREDIBLE INVENTION:

PURPOSE OF INCREDIBLE INVENTION:

ILLUSTRATION OF INCREDIBLE INVENTION:

NAME OF INCREDIBLE INVENTOR:

This invention is hereby granted a full patent by my lawyer,
Nigel, and if anyone steals it, then he will send them an
extremely angry letter.

ACKNOWLEDGEMENTS

* This book couldn't have happened without the following people.

† This book would have been rubbish without the following people.

‡ This book would have been exactly the same or maybe even slightly better without the following people.

My agents, Cath Summerhayes and Jess Cooper.*

My illustrator, Henry Paker.*/†

My husband, James.*/†

My editor, Ruth Knowles.*/†

My publishers, Francesca Dow and Tom Weldon.*

My editorial geniuses, Hannah Farrell and Justin Myers.†

My professors of publicity, Tania Vian-Smith and Dusty Miller.*

My designer, Jan Bielecki.*

My copy-editor, Wendy Shakespeare.†

My BFFs, Ruby and Ziggy.‡

My too many nephews and nieces: Noah, Zareen, Lenny, Sidney, Quinn, Jesse and Olive.‡

INDEX

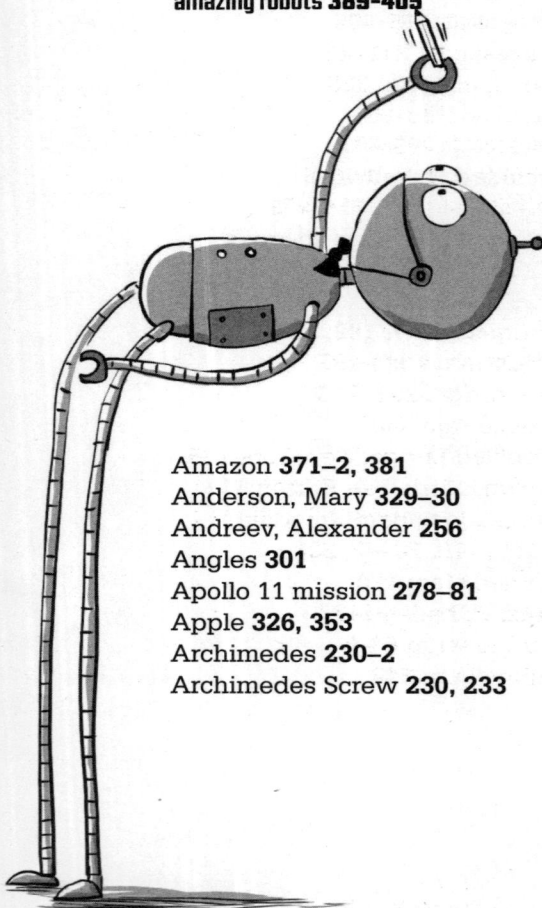

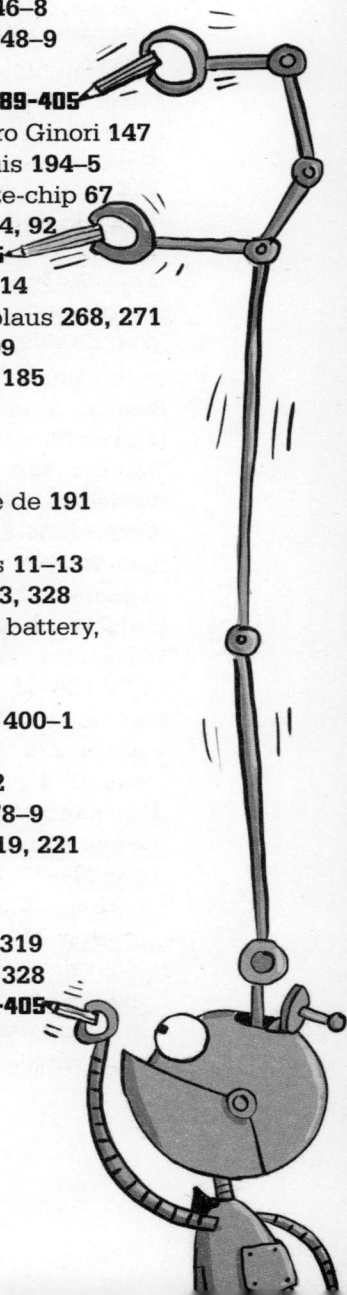

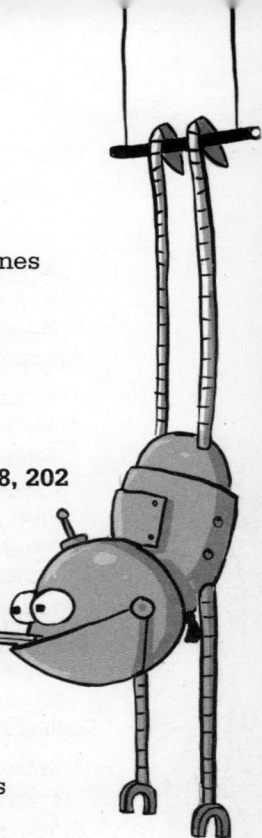

CREDITS

MANAGEMENT
Francesca Dow
Tom Weldon

EDITORIAL
Sarah Connelly
James Kay
Ruth Knowles
Philippa Neville

DESIGN & ART
Nigel Baines
Jan Bielecki
Anna Billson
Jacqui McDonough
Dave Smith
Sophie Stericker

COPY-EDITOR
Wendy Shakespeare

PROOFREADERS
BUTLERTRON-6000
Steph Barrett
Wendy Shakespeare
Debs Warner
Mandy Wood

INDEXER
Hilary Bird

EDITORIAL CONSULTANTS
Hannah Farrell
Justin Myers

FACT CHECKER
Michael Ward

PRODUCTION
Shabana Cho
Naomi Green
Jamie Taylor

OPERATIONS
Róisín Duffy
Bella Haigh

AUDIO
James Keyte
Michael Pender
Tom Rowbotham
Chris Thompson

EBOOK
Koko Ekong
Jessica Dunn Gabrielle
Chaumeil Mercer
Anya Wallace-Cook

INVENTORY
Katherine Whelan

FINANCE
James McMullan
Lona Teixeira Stevens

CONTRACTS & LEGAL
Mary Fox
Annie Lachmansingh
Michael Peters

SALES
Kat Baker
Toni Budden
Kate Lamming
Michaela Lock
Geraldine McBride
Emma Richards
Minnie Tindall
Rozzie Todd

MARKETING & COMMUNICATIONS
James McParland
Dusty Miller
Sophia Pringle
Tania Vian-Smith

RIGHTS
Maeve Banham
Clare Braganza
Stella Dodwell
Susanne Evans
Jouda Fahari-Edine
Beth Fennell
Alice Grigg
Magdalena Morris
Anda Podaru

BIBLIOGRAPHIC METADATA
Charlotte Moore

PRINTERS
Julia Cassells
Adam Dyer
Antonia Key

ADAM'S LAWYER
Nigel Rosenkrantz KC

ADAM KAY used to be a doctor but is now a writer, which is good news for any readers who like his books and for patients everywhere.

#KaysIncredibleInventions

HENRY PAKER was once a little boy who did silly doodles in the margins of books. Now he is a grown man who does silly doodles in the middle of books.

THE COMPLETE AND COMPLETELY DISGUSTING COLLECTION OF BOOKS BY
ADAM KAY & HENRY PAKER

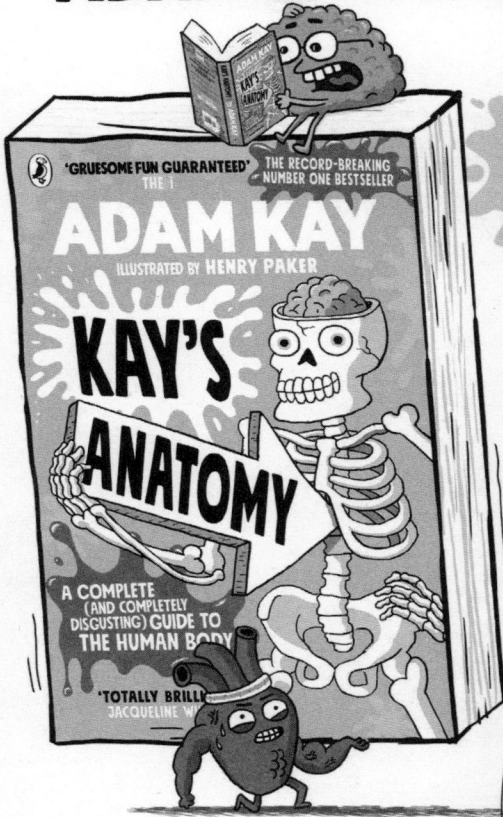

'HILARIOUS AND FASCINATING! I WISH ADAM HAD BEEN MY BIOLOGY TEACHER'
KONNIE HUQ

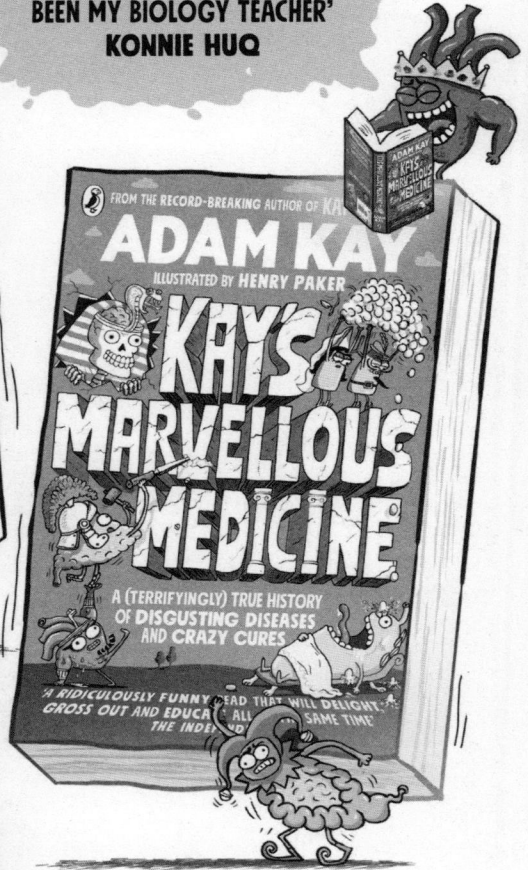

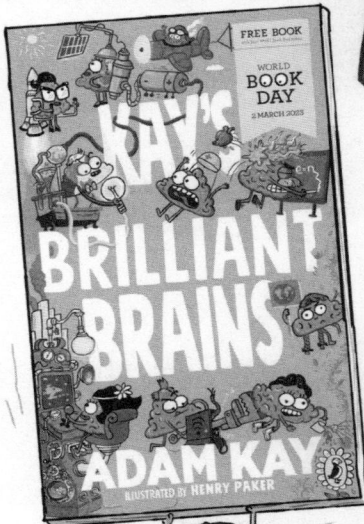

'GRUESOME FUN GUARANTEED' THE RECORD-BREAKING NUMBER ONE BESTSELLER
THE 1
ADAM KAY
ILLUSTRATED BY HENRY PAKER
KAY'S ANATOMY
A COMPLETE (AND COMPLETELY DISGUSTING) GUIDE TO THE HUMAN BODY
'TOTALLY BRILL...'
JACQUELINE WI...

FROM THE RECORD-BREAKING AUTHOR OF KAY...
ADAM KAY
ILLUSTRATED BY HENRY PAKER
KAY'S MARVELLOUS MEDICINE
A (TERRIFYINGLY) TRUE HISTORY OF DISGUSTING DISEASES AND CRAZY CURES
A RIDICULOUSLY FUNNY READ THAT WILL DELIGHT, GROSS OUT AND EDUCATE ALL... THE... SAME TIME'
THE INDE...

FREE BOOK
WORLD BOOK DAY
2 MARCH 2023

KAY'S BRILLIANT BRAINS
ADAM KAY
ILLUSTRATED BY HENRY PAKER

A WORLD BOOK DAY SPECIAL!

DONK!

"Wow! What is this place?" asked Amy.

"This is Noah's stomach," said a wise old raisin.

Amy noticed the baked bean sandwich Noah had eaten for breakfast.

Hey, don't judge him. We all like different things in our sandwiches.

Then
had ea

IMAGINE THIS
IN COLOUR . . .

chocolate cake and ketchup Noah
h. (OK, you can judge him for that.)

ADAM KAY

HENRY PAKER

FROM THE **BESTSELLING**
AUTHOR & ILLUSTRATOR
OF **KAY'S ANATOMY**

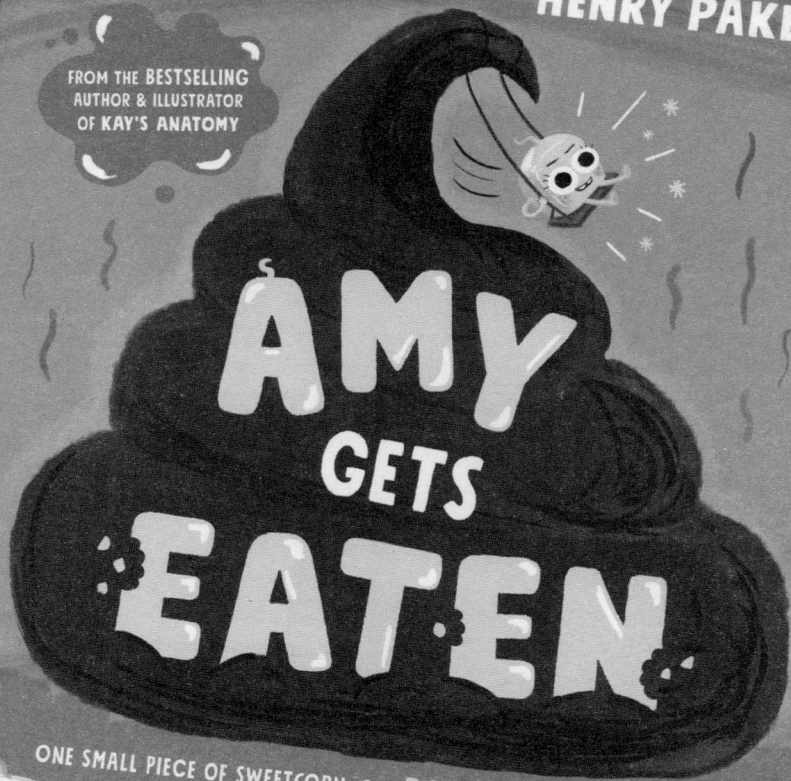

AMY
GETS
EATEN

ONE SMALL PIECE OF SWEETCORN. ONE **POO**-NORMOUS ADVENTURE

PLEASE IGNORE ME   I'M JUST A NORMAL TROUT